Garden Weeds

INKATA PRESS
London, Melbourne and Sydney

Published 1980

(c) R. J. Chancellor and H. R. Broad

ISBN 0 909605 21 1

Cataloguing in Publication Data

Chancellor, R. J. (Richard John)
  Garden weeds.

  Index
  Bibliography
  ISBN 0 909605 21 1

  1. Weed control. I. Broad, H. R. II. Title.

635'04958

Edited by Patricia Sellar

Cover design by Pauline McClenahan

Type set in Helvetica
by Savage & Co Pty Ltd Brisbane and
printed by The Dominion Press, Melbourne

Garden Weeds
and their control

Richard J. Chancellor

With illustrations by
Hilary Broad

INKATA PRESS

CONTENTS

INTRODUCTION

Much of the pleasure of gardening lies in the enjoyment of the flowers and vegetables that can be grown, but there is too a certain satisfaction in keeping the garden tidy and the weeds under control. Weeding is especially gratifying in that it provides immediate and obvious evidence of one's activity. Unfortunately, the evidence of this labour often disappears only too rapidly, for there are some weeds which are extremely persistent and reappear no matter how often or how carefully they are removed. The problem is usually complicated by the presence of delicate and carefully nurtured ornamentals or vegetables that have been hopefully planted in the same position, so the gardener is prevented from taking any really drastic action against the persistent weeds.

The purpose of this book is to try, in the light of the latest knowledge, to assist the gardener to suppress or actually eradicate the offending weeds within the existing garden layout. There are occasions, of course, when this is not practicable and one has to choose between sacrificing a cherished plant with a view to replanting later, or learning to live with a problem weed while keeping it within bounds. It is hoped that this book will help gardeners to make the right decisions, and help to minimise the labour involved in controlling their problem weeds.

Few gardeners are certain of the names of weeds or have knowledge of the characteristics that contribute to their unwelcome behaviour; even fewer gardeners know what can be achieved with herbicides, so I have aimed to make the book reasonably comprehensive and straightforward. As there is no other book available on this aspect of gardening, I have started by considering why weeds are so successful and discussing various aspects of weeds and their behaviour, to give a better understanding of the problems involved. The main part of the book consists of illustrations of the worst weeds of gardens, with accompanying notes, which should help with identification and with finding out how best to control them. Next, there is a summary of the different methods of controlling weeds, and then a discussion of the possibilities and problems of weed control in various situations in the garden. Finally, there is some advice about herbicides and their possible use. I have made every effort to ensure that all the facts and advice in this book are correct, but obviously no responsibility can be taken for any accident, loss or damage sustained through carrying out treatments recommended in this book as I have no control over their application. Nonetheless, I hope that the advice offered will always lead to a considerable reduction of weeds.

This book is written primarily for British gardeners; if used elsewhere, different conditions and different weeds may require different treatments. Similarly, regulations governing the use of herbicides may differ in other countries.

I should like to thank all those colleagues at the Weed Research Organisation, and others elsewhere, who spent so much time in reading the manuscript, and who offered useful comments and suggestions.

R.J.C.

1. THE PROBLEMS OF WEEDS

ANNUAL WEEDS

Weeds can be divided conveniently into annuals and perennials, as the two groups differ, often quite markedly, in their behaviour. Annuals complete their whole life-cycle within a period of twelve months, although not necessarily within a single calendar year, and so rely entirely upon seeds for their survival. In fact, annual weeds are perennial problems only because of their seeds. Annuals, with their short life-cycle, have a better chance of producing seed when growing in gardens than do biennials or perennials, most of which do not flower in the first year. Weed control in gardens, although possibly erratic, is certainly fairly frequent, so even the annual weeds found there will usually have short life-cycles. For most annual weeds in gardens, survival depends on the capacity to produce seed before the plant is destroyed.

The Number of Seeds in the Soil

As annuals rely absolutely on seeds, they must produce them in considerable numbers to allow for the odd mishap, such as an over-zealous gardener. As a result, the number of seeds in the soil can be remarkably high. The greatest number of seeds so far recorded from soil is about 350 million per acre, about 8000 per square foot or 56 per square inch! A useful measure of the number in your garden can be obtained by counting the seedlings in a small area; usually less than 10% of the seeds in the soil will germinate in any one year. Assuming that 10% (one in ten) germinate, and 100 seedlings are counted in an area of 1 square foot, then there are approximately 900 seeds in the soil beneath them. This works out at about 40 million to the acre, so in a garden of one-tenth of an acre there would be about four million seeds.

Dormancy in Weed Seeds

The fact that only a small proportion of the seeds germinate in any one year adds considerably to the weed's chances of survival. In most weeds, the germination of seed is regulated by dormancy. This is usually a physiological condition within the seed which requires the temperature and other factors to be appropriate before dormancy is ended and the seed germinates. As each seed differs to some extent in its requirements and in the environmental conditions to which it is subjected, germination of any weed will be distributed, not only over several months of the year, but also over several years. Weeds can, therefore, continue to exist over any period of time that is unsuitable for them to grow and set seed. This could be merely over a cold winter, or over a period of several years when, for example, that part of a garden is sown down to grass as a lawn.

Seasons when Weed Seeds Germinate

The dormancy-breaking requirements differ from one weed to another and this ensures that no two weeds germinate at exactly the same season of the year. Many do overlap, however, and the most usual times for weed seedling emergence are during spring and autumn – just, in fact, at the very time when you are sowing seeds of flowers and vegetables. This, of course, makes weed control more difficult, as flowers, vegetables and weeds will all come up together. There are a few weeds which germinate at other times, such as in mid-summer and in winter if it is not too cold, but these are not usually the worst weeds; some of the worst weeds will still germinate to a limited extent even during these less favourable seasons. It is likely that the longer the period during which its seeds will germinate, the more successful the weed.

If a weed has only a limited period of germination, then it can be controlled for a whole year by killing the seedlings that emerge during that limited period. For example, Red Dead-nettle (*Lamium purpureum*) germinates only in the spring, so a single cultivation in early summer can control this plant completely for the whole year.

A number of common garden weeds, such as Shepherd's-purse (*Capsella bursa-pastoris*), Groundsel (*Senecio vulgaris*), Annual Meadow-grass (*Poa annua*) and Chickweed (*Stellaria media*) can germinate during any month of the year, so constant vigilance is necessary to keep them under control.

The Length of Life of Seeds

As mentioned above, dormancy of the seed helps a weed to survive periods, often prolonged, which are unfavourable for growth of the plant. Quite distinct from dormancy is the length of life of the seed, which is also important. As every gardener knows, the seeds of some crops, such as the parsnip, do not last much longer than a single year, while others have a much longer life. To some extent, the larger the seed the longer it lives, so peas, beans and marrow seeds can remain alive for several years. In the case of weeds, this rule does not appear to apply as some of the smallest seeds have the longest lives. Certainly most weeds have seeds that are capable of living for five to ten years or even longer in the soil, although most of the seeds will probably survive for only a year or two. As it is the few that do survive that are important, studies have been made of the rate of loss of viability. The results show that under conditions of frequent cultivation a mixed population of weeds will decline by about a half in each twelve months, provided (and this is very important) that no fresh seed is shed during that time. With few cultivations, the decline will be slower, and with no cultivations at all the decline can be as little as 12% per year. So with frequent cultivations and with no fresh seeds being shed, a seed population can be halved every year. If you start with a population of 100 seeds in a patch of ground, after one year of effective weeding you will have about 50 seeds left and after two years only 25. If you continue for long enough, it will be seven years before your 100 seeds have been reduced to one, which confirms very nicely the old saying 'one year's seeding means seven years weeding'. If, as in the example above, you have 900 seeds per square foot, then it will take about ten years to reduce this to one seed to the square foot. In the garden of one-tenth of an acre, there will still be about 4000 seeds. Remember, this applies only if no new seeds have been shed during that ten years.

There are exceptions to every rule and a few weeds do have seeds that survive for only a short time in the soil, but these are in a minority. Most weeds have seeds that are capable of lasting for up to about ten years in the soil, and a few weeds have seeds capable of surviving for several decades or even longer. For example, it has been found that a few seeds of the Greater Plantain (*Plantago major*) survived 40 years burial in soil, and a few seeds of the Curled Dock (*Rumex crispus*) survived for 80 years; both of these plants can occur as garden weeds.

Prevent Seeding of Weeds
It is essential to prevent annual weeds from forming seed by destroying the plant before flowering. It must be before flowering, because some weeds can still produce seeds even if cut down during flowering and left lying rootless on the ground. These are ones that do not dry out quickly and, by retaining moisture in their stems, can continue to live for long enough to set seed. The immature seeds of certain weeds are capable of germinating. Weeds such as Groundsel (*Senecio vulgaris*), Annual Meadow-grass (*Poa annua*) and even Chickweed (*Stellaria media*) frequently produce seeds if cut down during flowering. If the weather is wet after they have been cut down, there is a good chance that they will re-root and continue to grow almost as if nothing had happened.

Methods of Invasion
Although a garden may be virtually weed-free, it can quickly become infested from outside unless vigilance is maintained. Wind-borne seed is a frequent source of trouble. Groundsel and various thistles as well as several perennial weeds often invade in this way, for, with the plume of hairs on their seeds, they catch the lightest wind and blow about over great distances. There is no way of preventing this invasion, so it is necessary to watch out for weeds that have jumped the garden fence.

Another source of infestation is soil brought in from outside. The friendly neighbour who has surplus seedlings can be giving you a problem as well as potential vegetables, for even a handful of soil can contain several hundred seeds. Even weed plants themselves can unwittingly be imported. The tell-tale shoots of Broad-leaved Willow-herb (*Epilobium montanum*) and the bright green rosettes of Hairy Bitter-cress (*Cardamine hirsuta*) are to be seen far too often on the soil of plastic containers sold at every garden centre. For every rosette there may also be a hundred seeds in the soil which cannot be seen.

Explosive Seed Dispersal
Although only one or two seeds or seedlings of a plant like Hairy Bitter-cress are introduced, it is able to disperse itself once it is inside the garden, for it has a vigorous seed-ejecting mechanism which can throw the seeds several feet from the parent. A woman gardener, who wrote to us once, was delighted with the light green rosettes and bright white flowers when she first saw this weed, and she actually transplanted seedlings to various other suitable places. Her delight soon turned to dismay when her garden became full of them. Be warned, for there are several garden weeds which eject their seeds in this way; they are, Annual Mercury (*Mercurialis annua*), Yellow-flowered Oxalis (*Oxalis corniculata*), Spring Beauty (*Montia perfoliata*) and Sun Spurge (*Euphorbia helioscopia*).

Three Generations a Year
The key to the success of annual weeds lies mainly in the large number of seeds that they can produce. There are, however, some that do not produce so many seeds, but increase their output in another way. Given reasonable weather, they can produce up to three successive generations in a single year and so increase their progeny that way. Annual Nettle (*Urtica urens*) and Groundsel (*Senecio vulgaris*) are good examples of this type of weed. These two also have the ability to produce seeds very early on in the life of the plant. This gives them an additional advantage over other weeds as they are more likely to produce seeds before they can be cut down or rooted out.

PERENNIAL WEEDS
Perennial weeds are different from annuals in many ways, apart from making a nuisance of themselves. They live for much longer; by definition, living for more than 24 months. In consequence, they have a different 'philosophy' to annuals, which are all for a quick turn round from seed to plant to more seed and plenty of it. Most perennials are slower, more easy-going and often, quite literally, creep up on you. They mostly rely on vegetative spread and many have far-creeping stems or roots, although seed may be produced also. Of the perennials listed in this book, about two-thirds are creeping plants and about three-quarters of these creeping plants produce seed regularly. However, even if they are able to produce seed, they may not necessarily start flowering for several years, especially if constantly checked by hoeing or cutting. Biennials also do not usually flower in their first year, but store up food in their roots for the great occasion in the second year. This makes them ineffectual as weeds, especially in gardens where they are noticed quickly, and destroyed before flowering. Perennials have an advantage over biennials and annuals in that they do not die after flowering, but may flower year after year.

Tap-rooted Perennials
These are plants that build up a large root filled with food reserves, like biennials, but differ from biennials as they can either throw up tall flowering shoots every year for many years like Docks (*Rumex* species), or grow a succession of separate flower heads like Dandelion (*Taraxacum officinale*). Cutting the plant down to ground level each year makes no difference, as it will flower again just the same. The plant must either be rooted out or treated with a herbicide. Both of the two common Docks can produce new shoots directly from the tap root, but only from the top 3-4 inches; if dug out to a greater depth, no new shoots will appear. However, the Dandelion can do better than that as it is able to produce shoots from any part of the tap root no matter where it is cut off; for the Dandelion at least, a herbicide is the easiest way, as it is susceptible to 2,4-D.

Weeds with Underground Creeping Stems
There are many weeds with underground creeping
stems. Underground stems do not grow permanently
underground, but turn up at the tip, usually after the
growing season is over in the autumn, or if frost sensitive,
in the following spring, to give rise to new aerial shoots

While the stem is below ground it is long, usually white,
leafless and generally straight, with nodes along its
length and a bud at each node. These buds do not
normally grow unless the plant is disturbed or the
underground stem cut up. When it is cut up, one or more
of the buds on each piece give rise to new aerial shoots
and hence to new plants. As every gardener knows, the
smallest fragment of an underground stem is apparently
capable of regrowth. Being brittle, these stems are easily
fragmented so that many pieces get left in the soil no
matter how carefully the ground is turned over.

Some of the worst creeping perennial weeds, such as
Ground Elder (*Aegopodium podagraria*), Couch-grass
(*Agropyron repens*) and Horsetail (*Equisetum arvense*),
which have underground creeping stems, occur mainly
in flower beds and other cultivated ground. These are
almost impossible to eradicate completely when growing
amongst ornamental plants, unless really drastic action
is taken. Creeping perennials can also occur in lawns.
Yarrow (*Achillea millefolium*), which also has creeping
underground stems, is one of the commonest. Some
weeds have surface creeping stems, such as Creeping
Buttercup (*Ranunculus repens*) and Slender Speedwell
(*Veronica filiformis*), both of which grow in flower and
vegetable beds as well as lawns.

Weeds with Underground Creeping Roots
Another group of problem perennials consists of those
with creeping roots. These have spreading root systems,
which arise as lateral branches from the vertical roots.
They stay permanently underground, although they do
produce aerial stems at intervals along their length, and
vertical roots too. Like creeping underground stems, they
can give rise to new plants if cut up by cultivation. There
are relatively few plants in this group, but they are some
of the worst weeds of gardens, for example, Creeping
Thistle (*Cirsium arvense*), Perennial Sow-thistle (*Sonchus
arvensis*) and Creeping Yellow-cress (*Rorippa sylvestris*).

Weeds Producing Bulbs
There are three bulb-producing weeds of importance in
gardens and these are possibly the most intractable
problems of all. They are the two Pink-flowered Oxalis
(*Oxalis latifolia* and *O. corymbosa*) and Lesser Celandine
(*Ranunculus ficaria*). There are stories of people selling
their houses because the garden was hopelessly infested
with one or other of the Pink-flowered Oxalis and they
may well be true. Certainly, when thoroughly established
in a garden there is no real remedy and as yet there is
no herbicide that will control these plants satisfactorily,
although one under development may be available later.
Cultivation is very rarely effective and usually serves
merely to spread the bulbs about and to encourage them
into further growth. So at the first sighting of the leaves,
which have three leaflets like Clover, take drastic action.
Do not allow the weed to become established in your
garden. Dig out the plant and the surrounding soil and
dispose of it, preferably by sterilization. Lesser Celandine
does not pose so terrible a problem, but nonetheless can
be a great nuisance; the small bulbils produced in the
angles of the leaves are shed easily and grow readily.

The Types of Problem
It can be seen therefore that the problems presented by
annual and perennial weeds differ greatly. Annuals
produce seed, usually in abundance and usually early on
in their life; much of the seed can remain dormant in the
soil for several years. Perennials, on the other hand, do
not necessarily produce much seed, some produce none,
but they have equally successful methods of
reproduction, such as creeping underground stems or
bulbils with which to spread themselves, often more
effectively than annuals. As described in Chapter 5, these
differences can affect the methods needed for
satisfactory control.

2. HOW TO IDENTIFY A WEED

The identification of weeds is an essential preliminary to practical weed control; once the enemy is known the battle is half won. Do not be put off by unpronounceable botanical names or if the common name doesn't always seem to be appropriate. First of all, look at the simple grouping system below. To help with identification, the weeds have been arranged in groups which are based on various characteristics. Find out to which group your weed belongs, and then go through the brief descriptions until you find the right one. Then check with the illustration to make sure. If you prefer to, just look straight through the pictures. The illustrations are arranged in a sequence based on leaf outline and not in the groups of the key; this is because some weeds have been included in more than one group to make the key easier.

If you do not find your weed at all, it could be that you have one that is not illustrated in this book. Do not be dismayed, for it is likely then that it will not be too serious a problem, but do seek expert advice. The weeds in this book have been very carefully selected as the ones most likely to be found as problems in gardens. Of course almost any weed or wild plant could occur there, but the ones illustrated are those that are most frequent and those which present the most intractable problems. To cover this last group adequately, one or two less common weeds have been included, such as Blue Sow-thistle and Creeping Bellflower. This is because they are such intolerable nuisances when they do occur that it is worth being warned against them. These plants were originally introduced as ornamentals because of their beauty, but quickly had their status changed to that of weed when their true nature was revealed.

When you have identified your weed, read about its characteristics and how to deal with it. Then turn to **Chapter 5 at the end of the book (page 79) for details of control methods.**

The Weeds in the Groups

Go through the groups in order and find the right one.
Then select the weed from the descriptions and check
it against the picture.

Group 1 *Grass weeds*

Page

COUCH-GRASS: Vigorous perennial grass with
creeping underground stems. — 78
BARREN BROME: Tall annual grass, softly hairy;
drooping flower head; seeds ½ inch long. — 77
ANNUAL MEADOW-GRASS: Small annual grass,
hairless; leaves short, curved, often wrinkled. — 76
(Other grasses may also occur, but are less serious
problems.)

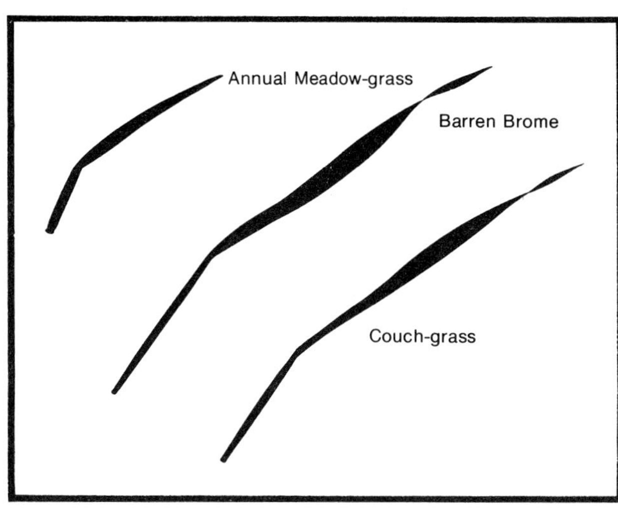

Group 2 *Perennial weeds with stems creeping along the
soil surface*

STINGING NETTLE: Leaves large and pointed
with stinging hairs; roots thick and yellow;
creeping stems purple. — 45
SLENDER SPEEDWELL: Stems many, thin and
bearing small kidney-shaped leaves; flowers
bright blue. — 53
WHITE CLOVER: Leaves divided into three
leaflets each marked with a white crescent;
flowers white or pink. — 25
YELLOW-FLOWERED OXALIS: Leaves divided
into three leaflets, often purple; flowers yellow. — 27
CREEPING BUTTERCUP: Leaves divided into
three leaflets, which are further lobed or
divided; flowers bright yellow; plant
far-creeping. — 29

Group 3 *Perennial weeds with stems creeping
underground (underground stems coloured)*

YARROW: Underground stems pinkish; leaves
much divided into narrow segments; flowers
white or pink. — 21
ENCHANTER'S NIGHTSHADE: Underground
stems white tinged with crimson, especially at
the nodes, extremely brittle; flowers white. — 64
HORSETAIL: Underground stems with small
tubers, both yellow when young but turning
black; aerial stems jointed, with many narrow
branches. — 20
JAPANESE KNOTWEED: Underground stems
light to dark rust-brown, stout; plant very tall;
flowers cream-coloured. — 68

Group 4 *Perennial weeds with creeping underground
stems or roots (underground parts white)*

CREEPING THISTLE: Leaves prickly with sharp
spines; flowers purple. — 40
HEDGE BINDWEED: Weed with twining stems;
tall rampant climber, mainly along hedges and
fences; flowers white or pink, bell-shaped. — 67
FIELD BINDWEED: Weed with twining stems;
climbing plant mostly in flower and vegetable
beds; flowers white or pink, bell-shaped. — 66
GROUND ELDER: Leaves divided into several
segments, each leaf on a single stalk arising
out of the ground; flowers white in flat heads. — 30
BLUE SOW-THISTLE: Leaves toothed,
bluish-green; flower stalks long, slender, tall,
bearing attractive light blue flowers. — 51
CREEPING BELLFLOWER: Flowers large,
purple, nodding; roots thick, like white carrots. — 61
WINTER HELIOTROPE: Leaves very large,
kidney-shaped; flowers light mauve,
vanilla-scented, winter-flowering. — 60
CREEPING YELLOW-CRESS: Leaves deeply
divided, often olive-coloured, forming small
rosettes; flowers yellow. — 31
PERENNIAL SOW-THISTLE: Leaves with large
teeth, bluish-green; flowering stems tall,
slender, sticky with glandular hairs; flowers
yellow. — 36
COLTSFOOT: Leaves very large, white with hairs
when young, angular when mature; flowers
yellow, appearing in early spring before the
leaves. — 46
HOARY CRESS: Leaves grey-green; flowers
white; creeping roots slender. — 59
ROSEBAY WILLOW-HERB: Stems tall, leafy;
leaves long, narrow, slightly toothed; flowers
bright purple. — 49
WHITE DEAD-NETTLE: Stems square; leaves in
opposite pairs, pointed, like nettle leaves;
flowers white. — 43

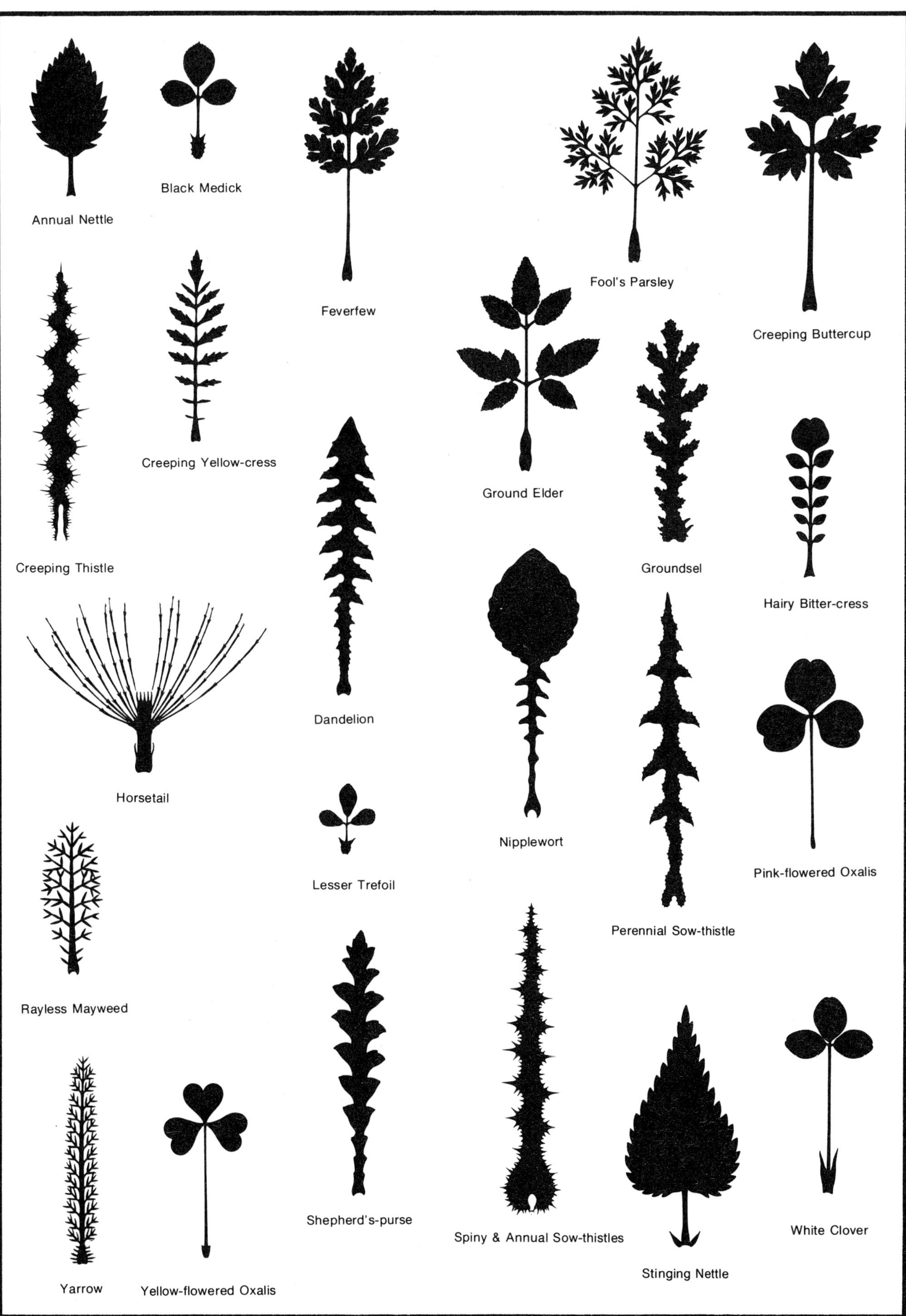

Annual Nettle
Black Medick
Feverfew
Fool's Parsley
Creeping Buttercup
Creeping Thistle
Creeping Yellow-cress
Ground Elder
Groundsel
Hairy Bitter-cress
Horsetail
Dandelion
Nipplewort
Pink-flowered Oxalis
Rayless Mayweed
Lesser Trefoil
Perennial Sow-thistle
Shepherd's-purse
Spiny & Annual Sow-thistles
Stinging Nettle
White Clover
Yarrow
Yellow-flowered Oxalis

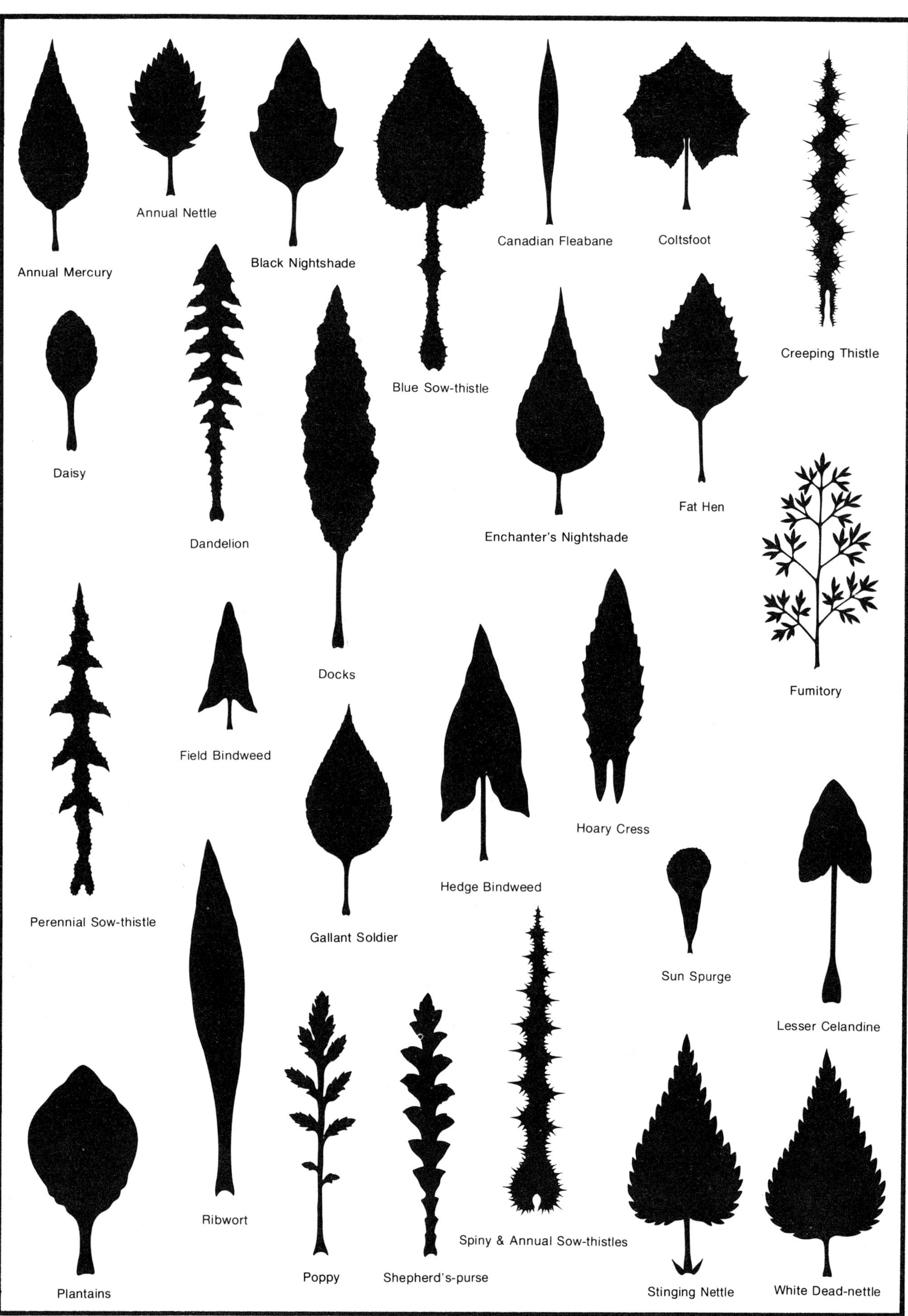

Annual Mercury
Annual Nettle
Black Nightshade
Canadian Fleabane
Coltsfoot
Creeping Thistle
Daisy
Dandelion
Blue Sow-thistle
Enchanter's Nightshade
Fat Hen
Fumitory
Docks
Field Bindweed
Hedge Bindweed
Hoary Cress
Sun Spurge
Lesser Celandine
Perennial Sow-thistle
Gallant Soldier
Plantains
Ribwort
Poppy
Shepherd's-purse
Spiny & Annual Sow-thistles
Stinging Nettle
White Dead-nettle

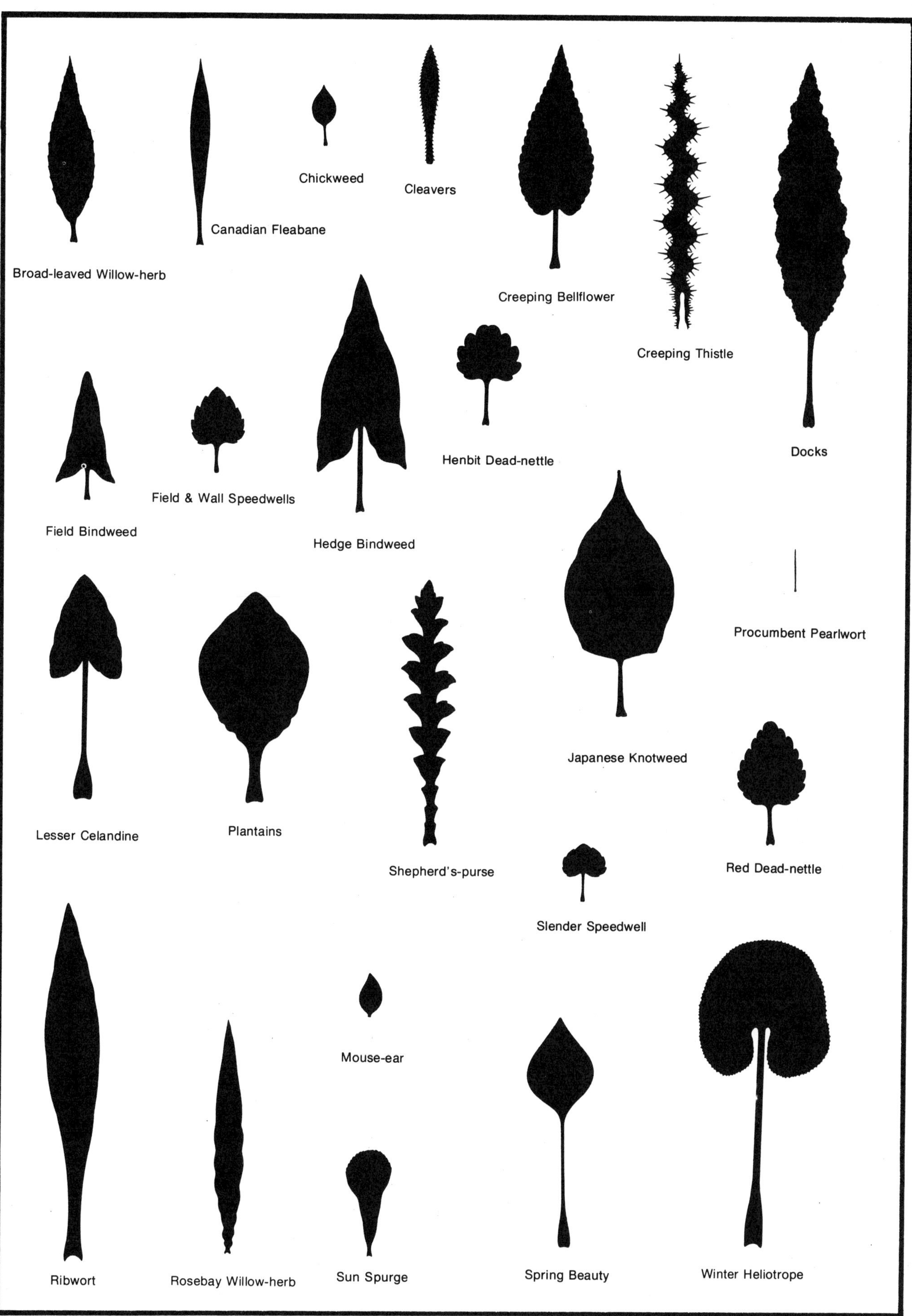

Chickweed
Cleavers
Canadian Fleabane
Broad-leaved Willow-herb
Creeping Bellflower
Creeping Thistle
Docks
Henbit Dead-nettle
Field & Wall Speedwells
Field Bindweed
Hedge Bindweed
Procumbent Pearlwort
Japanese Knotweed
Lesser Celandine
Plantains
Shepherd's-purse
Slender Speedwell
Red Dead-nettle
Mouse-ear
Ribwort
Rosebay Willow-herb
Sun Spurge
Spring Beauty
Winter Heliotrope

4. ILLUSTRATIONS AND DETAILS OF WEEDS

The illustrations are arranged in a continuous sequence based as far as possible, on leaf outline – starting with weeds having leaves that are deeply dissected into separate lobes or leaflets, and continuing with those having leaves that are shallowly toothed or indented or wavy. Next there are those with leaves that have entire margins and are without lobes or teeth, and ending with grasses. Unfortunately some weeds have a wide variety of leaf shape and could therefore appear in several places. Variability, where it does occur, is mentioned in the key and in the text, but be warned, especially of Shepherd's-purse. Wherever possible, weeds with similar leaves are placed near to one another, but because of this occasional variability you may not always be able to pick out a weed immediately. If you haven't used the key, go straight through the illustrations until you find the right one.

Opposite each picture there is a short text giving various important facts about the weed, especially those characteristics that make it worthy of inclusion in this book. The text starts with the common name, which follows in general the list of common names recently prepared for the Botanical Society of the British Isles by J. G. Dony, C. M. Rob and F. H. Perring, *English Names of Wild Flowers* (Butterworths, 1974). For those who are interested, the botanical name is also given. There is then a brief description of important characteristics to confirm identification, with emphasis given to the differences between that weed and other similar ones.

The habitats or situations in which it normally grows are listed, including those both outside and inside the garden. Soil preferences, if any, and the distribution in the British Isles are also given. The distribution has been extracted from *Atlas of the British Flora* edited by F. H. Perring and S. M. Walters and published by Thomas Nelson & Sons for the Botanical Society of the British Isles. Details of reproduction and spread of the weed and methods of controlling it are then given. Further information on methods of control can be found in Chapter 5 on page 79.

RAYLESS MAYWEED

Description

Rayless Mayweed (*Matricaria matricarioides*) is a small annual weed, which was introduced from north east Asia via North America. It only grows to a few inches tall, but is nonetheless a very successful weed. The leaves, which are finely divided, are strongly aromatic when crushed (from which it is occasionally called Pineappleweed). Like the rest of the *Compositae* family, it has many small florets grouped together in heads, but unlike the other Mayweeds it has no white, petal-like ray florets, hence the name 'Rayless' Mayweed.

Occurrence

Rayless Mayweed is often abundant, growing mainly on paths and drives in gardens, in waste places and especially in farm gateways. Although small it can be a considerable nuisance when abundant, for it is conspicuous and looks untidy.

Reproduction and Spread

Although introduced into Britain as recently as 1871, it has spread rapidly, probably because the seeds were carried about the country in mud on the wheels of cars. Today, it occurs throughout the British Isles and is found on most soils. The seeds are formed in abundance with an average of 7000 per plant. They germinate mainly in spring and early summer and the seedlings develop quickly. Flowering takes place in June and July and the seed heads ripen in succession. The seeds fall from the heads when the plant is blown by the wind or touched.

Control

Rayless Mayweed is mainly a weed of paths, drives and odd corners. If only a few are present, they are worth hand pulling, for the stems are tough and won't break. If there are many, they should be hoed or treated with herbicide. Hoeing should be done when the weed is young, for later the roots become difficult to cut. On paths, dichlobenil or simazine can be used, that is simazine alone to control germinating weeds, or mixed with other herbicides to kill established weeds as well. See under Annual Weeds in Cultivated Land on page 82 in Chapter 6 for weedkillers suitable for flower and vegetable beds.

HORSETAIL

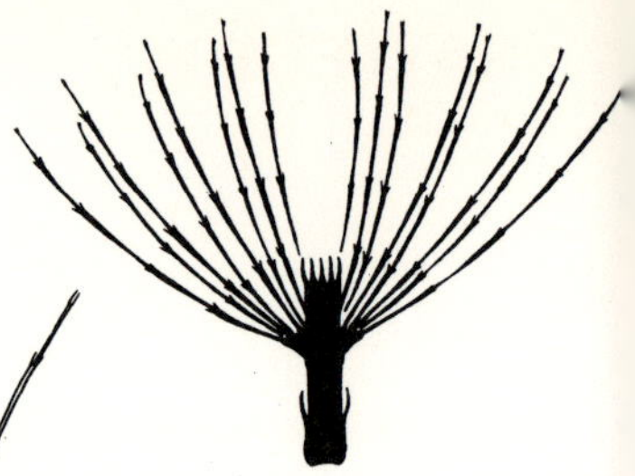

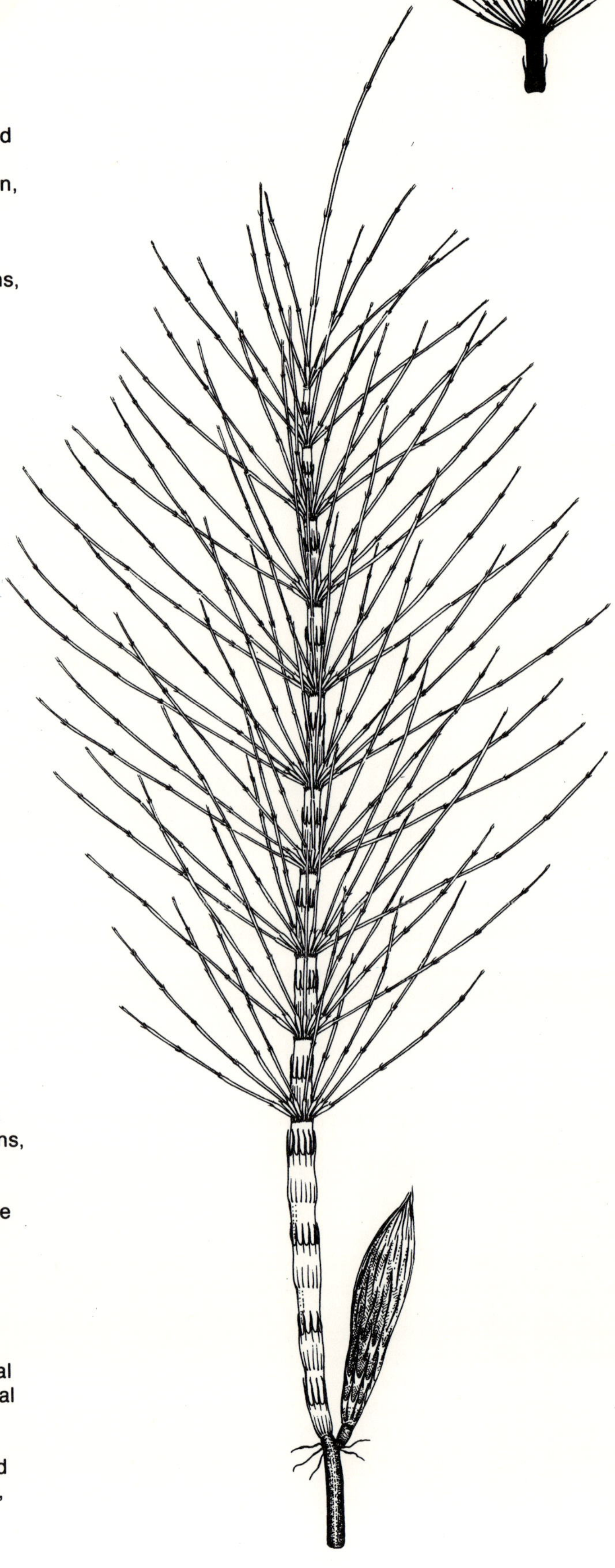

Description

Horsetail (*Equisetum arvense*) is a perennial weed with
short, much-branched, aerial shoots without proper
leaves. The thin branches are arranged in whorls around
the jointed stem; the whole resembles a horse's tail,
hence the common name. These stems, which are green,
are the sterile shoots; they emerge after the brownish
fertile ones. The fertile stems have a cone at the tip in
which spores are produced; no flowers are produced.
The aerial stems arise from creeping underground stems,
which are blackish and unlike those of other creeping
plants. The creeping stems also produce tubers. In
agriculture, horsetails are of especial significance
because they are poisonous, even when dried in hay.

Occurrence

Horsetail occurs in many situations – fields, hedges,
waste places and occasionally in gardens, where it can
be a serious problem. It grows on all soils throughout
the British Isles.

Reproduction and Spread

The inconspicuous, brownish, fertile stems emerge
above ground in March or April and produce a large
number of spores, which give rise eventually to new
plants, but only in moist situations. The fertile stems
quickly die back and are replaced by the more obvious
sterile shoots, which persist until the autumn. In gardens,
Horsetail is often a serious nuisance because of its
extensively creeping underground stems. Like most
creeping plants, fragments of these stems can generate
new plants, and so can the tubers that grow on them.

Control

Most herbicides will give only temporary kill of the aerial
roots, while the underground stems, which grow several
feet deep, can neither be dug out nor killed with
herbicides. Sodium chlorate will give some control,
where it can be used. On paths and around established
roses, many ornamentals, apples, pears, gooseberries,
raspberries, blackcurrants etc., dichlobenil can be
applied early in the year to prevent emergence.

YARROW

Description

Yarrow (*Achillea millefolium*) is a perennial weed, which grows to 18 inches (45 cm) tall when flowering. The flowers, which are pure white or less commonly pinkish in colour, are arranged in conspicuous flat-topped heads. Flowering is usually from June to September. The leaves are finely divided into a large number of narrow segments. It has creeping underground stems, which are pinkish in colour.

Occurrence

Yarrow is characteristic of grassland, and in gardens occurs almost only in lawns where it can be an unsightly nuisance. It grows on all soil types and is frequent and widespread throughout the British Isles.

Reproduction and Spread

Seedlings occasionally appear in cultivated soil, germinating both in spring and autumn, but they are never common, even though seed production can be prolific. The plant, which is often abundant in grassy places, spreads by means of slender, creeping, underground stems. The underground stems grow only slowly, up to 8 inches (20 cm) in a year, but they spread the plant very successfully. The tips of these stems turn upwards in the autumn and give rise to new rosettes of leaves around the parent plant. The creeping stems are capable of giving rise to new plants if cut into pieces.

Control

Control is not easy in lawns because the plant is relatively resistant to most weedkillers suitable for lawns. However, applications of lawn weedkillers containing dicamba, fenoprop, dichlorprop or mecoprop, which are sold mixed usually with MCPA or 2,4-D will give some measure of control; regular repeated treatments may be necessary to keep it in check.

FUMITORY

Description

Fumitory (*Fumaria officinalis*) is a not unattractive annual weed. The much-divided leaves are a waxy, light blue-green colour and the flowers, which are conspicuous and borne in spires, are a delicate pink with the tips and sides touched with dark purple. The plant is rather bushy in appearance, grows to about 18 inches (45 cm) tall and is short-lived.

Occurrence

Fumitory is of importance only when abundant. It grows in arable land, waste places and in gardens, and indeed in all cultivated ground. In gardens, it occurs mainly in new ones that have recently been divided up from farm land; it is more of a farm than a garden weed, although it can do well in both. It prefers lighter, poorer soils, both sandy and chalky, where it is often a serious weed. It occurs throughout the British Isles, but is most abundant in the east and south.

Reproduction and Spread

The seedlings emerge mainly in spring and early summer, but a few are to be found in late autumn. The plant flowers from May to September. Although it is often an abundant weed, it does not produce as much seed as many other weeds, for each flower produces only one. As a result it rarely becomes a problem in gardens. Nonetheless, its seeds can survive for up to 30 years in the soil, so it can be a persistent problem.

Control

As an annual weed confined to cultivated soil, it can be controlled either by hoeing or herbicides. The young plants are easily killed by hoeing and they can be hand pulled when necessary if grasped low down, but in larger numbers over a wider area, herbicides will need to be used. For suitable herbicides see under Annual Weeds in Cultivated Land on page 82 in Chapter 6 at the end of the book, and in the table (page 85).

BLACK MEDICK

Description

Black Medick (*Medicago lupulina*) is an annual or short-lived perennial weed, which grows flat along the ground or scrambles over other plants. The leaves are divided into three small leaflets. The flowers are bright yellow and give rise to coiled one-seeded pods, which are black when ripe. The small point at the end of the central vein of the leaflets distinguishes it from Lesser Trefoil, which has none. Frequently they both occupy the same situations and are very similar in appearance.

Occurrence

In gardens, this weed is a nuisance in lawns and rough grass, but can also occur in flower and vegetable beds and even on paths and drives. It is common in grassy places throughout the British Isles, although it is less widespread in Scotland and Northern Ireland.

Reproduction and Spread

The seeds germinate in both spring and autumn. The young plants grow rapidly and flower from April to August, when they become very noticeable. Mowing the lawn when the plants are seeding simply spreads the weed further and further over the lawn. The seeds can survive for several years in the soil.

Control

This is a very difficult weed to get rid of completely. At one time the only method of control was to cut through the main tap root with a sharp knife, but it is tough and wiry, which also makes it hard to pull by hand. Fortunately the herbicides dicamba, fenoprop, mecoprop and dichlorprop will give reasonable control and can be used on grass and lawns. They are usually sold mixed with MCPA or 2,4-D for lawns. As seeds survive for several years in the soil, a fresh crop of seedlings will appear each year, so that treatment will need to be repeated at least annually.

LESSER TREFOIL

Description

Lesser Trefoil (*Trifolium dubium*) is an annual weed with stems that lie along the soil surface. The leaves are composed of three leaflets. The plant closely resembles Black Medick, but the leaflets do not have a small projecting point at the end of the central vein and the seed pods are brownish and not black when ripe. The flowers are yellow.

Occurrence

This weed is very common in lawns where it can be seen as large patches of rather light green leaflets with numerous heads of small yellow flowers scattered among them.

Reproduction and Spread

The seeds germinate in spring and autumn and the young plants grow rapidly. The stems extend outwards to form mats, which under good conditions can be quite large. Flowering is from May to October and many seeds are produced.

Control

Lesser Trefoil, like White Clover, can be treated with dicamba, fenoprop, mecoprop or dichlorprop, but it is less responsive to these chemicals so several applications may be necessary. These chemicals are usually sold mixed with MCPA or 2,4-D for lawns in order to increase the range of weeds killed. Herbicide treatments should be made annually as fresh crops of seedlings will appear each year. As this is an annual weed and dependent entirely upon a central tap root, it is possible, when only a few plants are present, to cut this root and pull out the plants individually; each plant may cover a considerable area. It is of course essential to prevent any seed being shed by this weed. Mowing when the plants are seeding will spread the weed about the lawn.

WHITE CLOVER

Description

White Clover (*Trifolium repens*) is a surface creeping perennial weed with marked vegetative spread. The creeping stems root at the nodes. The leaves are divided into three leaflets, each of which has a white crescent-shaped marking across the base. The flowers, which are white or pinkish, are borne in clustered heads on short stalks.

Occurrence

White Clover is a frequent weed in lawns and can also occur elsewhere in gardens. The dark green leaves, which occur in patches on many an otherwise well-kept lawn, are very noticeable and consequently unsightly. It occurs throughout the British Isles.

Reproduction and Spread

Although White Clover produces seed, some of which can survive for up to twenty years in the soil, it relies much more on vegetative spread. The stems creep along the surface of the ground rooting at those nodes that touch the soil. It can with time form very large patches. It is suspected that most if not all of these usually originate from the time the lawn was sown.

Control

Repeated mowing of White Clover has no apparent effect although it may slow down the rate of spread of the plant. The leaves are possibly not quite so large afterwards, but they are rapidly replaced and certainly just as obvious. The only effective treatment in lawns is to spray the patches with dicamba, fenoprop, mecoprop or dichlorprop, which should be applied in late spring when growth is rapid. Several applications may be necessary. The weed is more or less unaffected by 2,4-D and MCPA, but these are included in many lawn weedkillers in order to control other weeds.

PINK-FLOWERED OXALIS

Description

Pink-flowered Oxalis (*Oxalis corymbosa* and *O. latifolia*)
are small attractive perennial plants between 6 inches
and 1 foot (15-30 cm) tall. The leaves are divided into
three heart-shaped leaflets. These weeds are
characterised by the production underground of a very
large number of small pinkish brown bulbils. Both plants
have pink flowers, but the leaves of *O. corymbosa* are
rather hairy and have a line of raised red dots along the
margins underneath, while those of *O. latifolia* do not.
O. latifolia also differs in having the bulbils on stalks.
O. corniculata produces no bulbils and has yellow flowers.

Occurrence

Both these plants were originally introduced as
ornamentals. They occur in market gardens, nurseries,
old gardens and in glasshouses, in all of which they can
be serious weeds. They are quite widespread in central
and southern counties of England and are especially
frequent either near London or in the south west.

Reproduction and Spread

Although they flower between July and September, no
seeds are formed. Once introduced into a garden, they
are very persistent indeed and the numerous bulbils are
quickly dispersed about the garden and into other
gardens too. The bulbils can remain dormant for long
periods in the soil and can withstand extremes of drought
and cold.

Control

Control of the weed in a garden is virtually impossible,
although a few gardeners have claimed to have
eradicated them by persistent weeding. Leaf destruction
and soil disturbance encourage bulbils to ripen and
cultivation spreads them about. Bulbils, if dug up, should
always be burnt. Infested areas can be grassed down for
several years, or the soil can be chemically fumigated,
or young pigs, if allowed to root in the earth, will eat and
destroy the bulbils; but none of these remedies are very
suitable for gardens! See the table on page 85 at the end
of the book.

YELLOW-FLOWERED OXALIS

Description

Yellow-flowered Oxalis (*Oxalis corniculata*) is a prostrate annual or perennial plant. The stems creep along the soil surface and root at the nodes. The leaves are divided into three leaflets, which are heart-shaped and vary greatly in size and colour. The flowers are yellow. There is another yellow-flowered Oxalis (*O. europaea*), which is only very rarely a weed in gardens. It differs in having underground creeping stems. *O. corniculata* was originally introduced. Many forms (it is very variable) have been cultivated as ornamentals, including purple-leaved and dwarf ones. These forms vary in frost hardiness; the susceptible ones behave as annuals, while the hardier ones survive the winter and are perennial.

Occurrence

As an ornamental or a weed, it is found most frequently in gardens. It occurs in lawns, glasshouses, flower beds, waste places, paths etc., and has probably been spread from garden to garden by the movement of pot plants and transplanted ornamentals. It grows scattered throughout the southern half of England and Wales.

Reproduction and Spread

Although the plant does extend itself by the creeping stems, it is by seed that it is spread. The seeds are forcibly flung out when ripe to a distance of up to 3 feet (1 metre) from the parent, which accounts for its ubiquity where it does occur. Flowering is from June to September.

Control

Although a troublesome weed on occasion, it is not nearly so serious as the two Pink-flowered Oxalis which produce bulbils. However, control is often difficult. The only method is to kill the plant before it seeds. On paths, paraquat/diquat, dichlobenil or mixtures containing simazine can be used, while on lawns, most lawn weedkillers will give some measure of control; but reinfestation is likely from seeds in the soil. For flower beds, etc., see under Annual Weeds in Cultivated Land (on page 82) at the end of the book, and in the table on page 85.

HAIRY BITTER-CRESS

Description

Hairy Bitter-cress (*Cardamine hirsuta*) is a very small
annual weed. The flowering stalks with their bright white
flowers reach a maximum height of only 9 inches (23 cm).
The leaves are divided into small, almost-circular leaflets;
they are arranged in up to seven opposite pairs with a
single terminal leaflet, which is kidney-shaped. In open
situations, the plant forms neat compact little rosettes,
which are not unattractive.

Occurrence

This weed occurs throughout the British Isles, but is only
really of importance in gardens and horticultural
nurseries where it can be a persistent nuisance. It is very
common in the plastic pots of many garden centres. Its
success as a weed is due mainly to its short life-cycle
and efficient method of seed dispersal.

Reproduction and Spread

The seeds germinate in spring and autumn and a rosette
is quickly formed. The brevity of the life-cycle allows two
or three generations a year and, as an average plant
produces about 600 seeds, this results in a rapid increase
of plant numbers. When the seed pod is ripe, it splits
open violently and flings out the seeds. In still air, they
can travel up to 3 feet (1 metre), and further when it is
windy.

Control

It is essential never to ignore a single growing plant of
Hairy Bitter-cress, pull it up or hoe it out at once! If
flowering, destroy it by burning. Never let it ripen seed,
for to touch it at that stage will set off the violent seed
dispersal. On paths, rosebeds and around many shrubs,
simazine or dichlobenil can be used to prevent
establishment or, if already established, then the mixture
of paraquat and diquat will give a speedy kill. For paths,
simazine often has other chemicals added to give cure
as well as prevention. Among flowers and certain
vegetables, propachlor can be used to prevent the weed
emerging. Hoeing is also very effective.

CREEPING BUTTERCUP

Description

Creeping Buttercup (*Ranunculus repens*) is a perennial weed with creeping stems which run along the surface of the soil forming rosettes of leaves at intervals. There are several thick white roots under each rosette. The flowering stems grow up to 2 feet (60 cm) tall and bear the glossy golden-yellow flowers which are characteristic of buttercups. There are three common buttercups which are very similar, but Creeping Buttercup, which alone is a problem in gardens, is the only one with creeping stems.

Occurrence

Creeping Buttercup is often a weed in gardens where, because it creeps rapidly about and germinates in abundance from seed, it can be very troublesome. It grows on cultivated soil and grassland and in many other situations too and occurs on most soils. In some situations, such as grassland, it tends to prefer damp soils, but elsewhere seems to like drier conditions too. It is a common plant throughout the British Isles.

Reproduction and Spread

It is the last of the three common buttercups to come into flower, from May to June. Up to 150 seeds are produced per plant. These are shed around the parent or are sometimes distributed by birds, and they are known to survive for several decades in the soil. The seeds germinate from March to November, and the seedlings develop quickly. In May and June, creeping stems are produced which run along the soil surface and root at the nodes. These form new plants. By this means, a single plant can colonise up to 40 square feet (4 sq. metres) in a single year. In gardens, these creeping stems are the greater nuisance.

Control

This plant can be hoed out easily when small; but when established, with its thick strong roots, it often needs digging out. It can never be pulled out by hand, although the runners will sometimes come up when very young. Fortunately, it is very susceptible to 2,4-D and MCPA, so mixtures containing either are a very effective method of control in lawns, where it can form much of the sward. On paths, it can be readily controlled by paraquat/diquat.

GROUND ELDER

Description

Ground Elder (*Aegopodium podagraria*), a perennial
weed, is also known as Goutweed, Bishops Weed and
Herb Gerard. It appears mainly as a mass of leaves
growing straight out of the ground; apart from the
flowering shoots, which are rarely seen in gardens
anyway, all the stems remain permanently below ground.
The leaves superficially resemble those of the Elder,
hence the common name. The underground creeping
stems are white and creep far and wide giving rise to
new leaves at every node. The flowering stems are erect
and bear heads of white flowers.

Occurrence

Although it was introduced
(probably deliberately by the
Romans for its supposed
medicinal value) and despite it
being confined to gardens, it is
found throughout the whole of the
British Isles at the present time.

Reproduction and Spread

The seeds germinate in spring,
but seedlings are rarely seen
because the plant is usually not
allowed to flower in gardens.
Therefore it relies almost entirely
upon vegetative spread. When
flowering does occur, it is from
June to August. Leaves emerge
above ground in February and die
back again in late autumn. They
arise from the underground
stems, which grow only a few
inches below the soil surface and
can creep up to 3 feet (90 cm) in
a year. They are brittle and pieces
left behind in the soil after digging
give rise to new plants.

Control

Control is very difficult because the creeping stems all too often grow
in among the roots of cherished ornamentals and cannot be
dislodged. If the ornamentals can be lifted and thorough cultivations
carried out, then with care the weed can be eradicated; when this
is not possible, the only remedy is persistent hoeing between plants
so as to cut off all leaves as soon as they appear, although this is
unlikely to kill the plant completely. Where sodium chlorate and
2,4,5-T can be used (see table on page 85 and Chapter 9 at the
end of the book), they will give useful control or even kill. Dichlobenil
can be used to prevent emergence around many woody ornamentals
and fruit bushes, provided that they are established, but it will need
to be re-applied every year.

CREEPING YELLOW-CRESS

Description

Creeping Yellow-cress (*Rorippa sylvestris*) is a small and relatively insignificant perennial weed, although the bright yellow flowers are conspicuous when open. It has a mass of thin creeping roots, which readily give rise to many new aerial shoots. The shoots bear a number of deeply-divided leaves, which in winter are very inconspicuous against the soil because they are then an olive green colour.

Occurrence

This plant, which occurs mainly in gardens and horticultural nurseries, is a very persistent and virtually ineradicable pest, especially when growing among ornamentals. It tends to prefer lighter soils, where it can become rampant. Although it grows throughout the whole British Isles, it is only scattered in occurrence as yet, especially in the north and west, but it is spreading.

Reproduction and Spread

The plant overwinters as inconspicuous, olive-coloured, leafy shoots, but in spring the new leaves are a brighter green. Flowering is between May and August and the bright yellow flowers are produced in profusion, but fortunately they are more or less sterile and little seed is formed. However, propagation by the creeping roots is both effective and rapid and, if attempts are made to dig it up, any small pieces left behind can produce new plants. Accidentally transplanting its roots about the garden or in other gardens should be avoided at all costs.

Control

There is no certain means of control. Constant hoeing and herbicides will keep it in check, but it is far better to prevent it from getting well established in the first place. If it is present already, prevent it from spreading – if necessary dig a trench. Its response to herbicides is largely unknown, but MCPA, 2,4-D or others might give some control, but remember they will kill other plants too! Dichlobenil could also be useful (see page 89 in Chapter 9). It is better to be ruthless than try to live with this weed.

GROUNDSEL

Description

Groundsel (*Senecio vulgaris*) is generally small, but always an untidy-looking weed, which grows to about 15 inches (40 cm) tall. The stems are succulent, soft and more or less erect. The leaves are ragged and coarsely toothed. The inconspicuous florets are yellow and grouped into heads.

Occurrence

As well as being unsightly, Groundsel is one of the commonest weeds and so can be a great nuisance, although not growing to any great size. Besides gardens, it grows on farmland, waste places, stone walls, roadsides and in many other situations. It occurs on all soil types and is more or less ubiquitous throughout the British Isles.

Reproduction and Spread

The seeds can germinate at any time of the year, and under good conditions the plant has a short life-cycle. As it can flower in all months, it is capable of producing three generations in a year. Each seed has a plume of hairs to catch the wind and, as the seeds are light, they can be carried over great distances. They are said to be capable of surviving for several years in the soil.

Control

Although Groundsel is a small plant, control is essential, but difficult. Hoeing is effective, especially for seedlings, but as seed can blow continuously over the garden fence this is likely to be a continuous process. Hand pulling of plants is not always successful, for the stem usually breaks and will regrow from the base unless dug out. If older plants are uprooted and left on the ground, they will take root again in suitable weather. In cultivated soil, propachlor, 2,4-DES and chloroxuron can be used (see under Annual Weeds in Cultivated Land (on page 82) at the end of the book). On paths, simazine or dichlobenil can be used.

NIPPLEWORT

Description

Nipplewort (*Lapsana communis*) is a tall annual weed growing to about 4 feet (120 cm). It forms a loose rosette of leaves from which the tall, slender, flowering stem grows. The stem bears small flower heads, which contain the clustered pale yellow florets. The lower leaves have a large terminal lobe and may have several small ones on either side of the stalk as well. The upper leaves are not divided.

Occurrence

Nipplewort is a very widespread and common weed, which grows in cultivated land, hedgerows and waste land. Although occurring also in farmland, it is much more a weed of gardens and waste places, where it is often frequent and always noticeable because of its height. It grows on all types of soil and occurs throughout the British Isles. It was formerly used as a salad plant, despite its bitter flavour, and this may partly account for its frequency.

Reproduction and Spread

The seeds germinate in both spring and autumn and subsequent growth is rapid. The plant flowers between June and September and on average 1000 seeds are produced by each plant. The plant has, however, one weak point, which is that the seeds are not shed until long after they are ripe, unless the plant is disturbed. By careful weeding, seed shed can be completely prevented even at a late stage. However, once the seeds are shed they can survive in the soil for several years, so it is essential to prevent shedding.

Control

The plant can readily be pulled by hand, if grasped low enough, and it is easy to kill by hoeing; as it has no tough roots, it cannot regrow from them. Paraquat/diquat and other chemicals suitable for cultivated soils will give good control, (see under Annual Weeds in Cultivated Land on page 82 at the end of the book, and in the table).

FOOL'S PARSLEY

Description

Fool's Parsley (*Aethusa cynapium*) is a small annual weed with divided dark green leaves, and small white flowers which are arranged in a flat head. It can be identified when young by the flat parsley-like leaves and, when flowering, by the long downward-pointing green bracts which fringe the edge of the flowering head. When flowering is over, the large whitish seeds are characteristic, for they remain on the head for a time before being shed. The plant is poisonous to some extent, and people have been poisoned when they mistook the leaves for parsley or the roots for radishes. This is surprising for it has an unpleasant smell.

Occurrence

Fool's Parsley occurs in cultivated land on farms and in gardens. It is common in central, eastern and southern England, but less so in the west and north. Although widespread and common, it is rarely a problem in gardens. It grows on most types of soil.

Reproduction and Spread

The seedlings emerge between March and May, although a few appear in late summer. Flowering takes place in July and August and the seeds are shed later on in the autumn. Up to 6000 seeds are produced per plant, and they can survive in the soil for up to ten years.

Control

This annual weed is confined to flower and vegetable beds and so can be hoed out or pulled by hand if only a few are present, but care should be taken to grasp the stem low down or else it will break. If a herbicide is required, dichlobenil, where it can be used, will prevent emergence and paraquat/diquat will kill seedlings; but chloroxuron is ineffective and simazine is not very effective. See at the end of the book (page 85) for control methods.

SHEPHERD'S-PURSE

Description

Shepherd's-purse (*Capsella bursa-pastoris*) is a small annual or biennial weed growing to about 18 inches (45 cm) tall. It is an extremely variable plant. The leaves can be entire in outline or show all degrees of dissection to an extreme herring-bone appearance. The leaves are generally hairy with very small star-like hairs. The plant forms a rosette of leaves, which give rise to the flowering stem. The flowers are small and white and the seed pods are triangular.

Occurrence

Shepherd's-purse is one of the most successful weeds and occurs in farm crops and in gardens, where it is widespread and common. It grows on all soil types and occurs throughout the British Isles.

Reproduction and Spread

The seeds can germinate in almost any month, although they do so mainly in spring and autumn. The early-appearing seedlings grow very rapidly and the whole life-cycle may take only six weeks; consequently, there may be two or three generations in a single year. Later-emerging seedlings overwinter as rosettes and can withstand even severe frost. The plant can flower and ripen seed in any month. Up to 4000 seeds are produced per plant. When wet, they are sticky and may be carried about on boots and garden tools. The seeds can survive in the soil for up to 30 years.

Control

The essence of control of this weed is to prevent seeding. As it is mainly a weed of cultivated soil, hoeing is very effective, but the roots are quite tough. When flowering, it can be pulled if the soil is moist. Various herbicides can be used to control it, although paraquat/diquat is not very effective. See the table on page 85 at the end of the book.

PERENNIAL SOW-THISTLE

Description

Perennial Sow-thistle (*Sonchus arvensis*) is a tall perennial weed growing up to 4 feet (120 cm) tall. The leaves are toothed and blue-green in colour. The tall, very erect and relatively little-branched flowering stems bear large flowers of a bright golden yellow. The stems are usually densely covered in yellow glandular hairs, which distinguish this plant from the two annual Sow-thistles. The plant produces many horizontal creeping roots, which give rise to new aerial stems at a distance from the parent; as a result, the plant often forms extensive patches. The stems exude a milky juice when broken.

Occurrence

Perennial Sow-thistle grows in all soils and is an especial nuisance when growing in heavy ones. It occurs in arable land, roadsides, gardens and other places throughout the British Isles, but it is less widespread in Scotland.

Reproduction and Spread

Established plants produce new aerial shoots above ground in April and May, forming at first a rosette of leaves and then a flowering shoot. The flowers open from July to October and many seeds are produced, up to 19 000 per plant have been recorded. The light seeds are dispersed by the wind over great distances. The creeping roots readily grow new shoots and even small fragments will do so if broken up by cultivation.

Control

Control in gardens is very difficult because of the extensive root system. When attempting to dig it up, all pieces must be removed, but this may be impossible as the roots can penetrate to a depth of 6 feet. As it occurs mainly among ornamentals and vegetables, application of 2,4-D or MCPA to the shoots with a paintbrush is often the only remedy. This treatment may need to be repeated, but keep it off desirable plants. Dichlobenil, where it can be used, also gives control. Watch should be kept for seedlings becoming established from seed blowing in from outside. See at the end of the book (page 85) for details of control methods.

POPPY

Description

Poppy (*Papaver rhoeas*) with the large scarlet flowers needs little description, for decorative selections of this plant are widely sown in gardens under the name of Shirley Poppy. The leaves vary in shape, but are all deeply lobed and hairy. Another poppy, which also occurs occasionally where it is not wanted, is the Opium Poppy (*P. somniferum*). It has white or purple flowers.

Occurrence

Poppies are very frequent in farmland, and also in gardens where they occur quite often from seeds shed by plants that were originally sown. Poppy occurs throughout the British Isles, but is most frequent in England. Opium Poppy is similar in distribution, but very much less common.

Reproduction and Spread

Poppies of all types produce a large amount of seed, and seed of the ordinary Poppy is capable of surviving in the soil for any time up to about 80 years. Therefore, once seed has been shed on to the soil one may expect seedlings to occur for many years afterwards.

Control

As it grows in cultivated soil, this weed (or ornamental) is easy to control. Hoeing is very effective, for the plant is soft and quickly dries up. When seedlings recur persistently, they may be worth leaving for their flowers, but pull them out before they seed. If herbicides are needed, Poppy is controlled by chloroxuron, chlorpropham and the paraquat/diquat mixture.

DANDELION

Description

Dandelion (*Taraxacum officinale*) is a perennial weed with a large thick tap root, which is crowned with a rosette of coarsely-toothed leaves. The bright yellow flowers, which must be familiar to everyone, arise on short, hollow stalks which have a staining juice.

Occurrence

Dandelion is one of the commonest garden weeds and grows in most situations, such as lawns, flower and vegetable beds, dry walls, drives and paths (as it is resistant to trampling) and in many other places. It grows on all soils and is abundant throughout the British Isles.

Reproduction and Spread

Dandelions flower from March to October and are particularly noticeable on roadsides in early spring. Seeds are produced profusely; over 5000 per plant each year have been recorded. Each seed has a large plume of hairs, which results in an almost unlimited dispersal by the wind. The roots are also very regenerative. It has been found that pieces taken from any part of the tap root are capable of producing new plants very rapidly. This contrasts with the tap roots of Docks, which can regenerate only from the top few inches.

Control

Regeneration from the roots makes hoeing a waste of time or at best only a temporary palliative. In flower beds and other cultivated ground where desirable plants are growing, it is best to apply a little 2,4-D with other chemicals (as for lawns) on to the leaves with a paintbrush, but avoid ornamentals and vegetables. Dichlobenil granules can be used under certain ornamental shrubs and fruit bushes, provided that they are established. In lawns, 2,4-D can be applied overall as the grass will not be harmed and other weeds will be controlled as well; more than one application may be necessary.

FEVERFEW

Description

Feverfew (*Chrysanthemum parthenium*) is a perennial weed which grows to 1-2 feet (30-60 cm) in height. The leaves are broad and deeply divided into separate lobes. The leaves are often softly hairy and have a characteristic strong scent when crushed. The flowers are like those of Daisies, with short, stubby white petals and a broad yellow centre.

Occurrence

Feverfew is usually found growing out of walls and in other dry places, but it can also be found occasionally as a weed in gardens, especially in cultivated soil. Its association with gardens may well be due to its having formerly been grown there for its supposed medicinal properties. It occurs throughout the British Isles.

Reproduction and Spread

Feverfew produces many seeds but, as they are not wind-dispersed, the plant tends to be very local in its distribution. Flowering occurs in July and August and the seedlings appear in autumn and spring.

Control

Feverfew never occurs in large numbers, so the seedlings, which are small, can easily be hoed out; if present as large plants, they can be pulled out by hand, which leaves their pleasant scent on the hands. This weed rarely causes much trouble, but its extensive root system will compete with neighbouring plants. Its response to herbicides is not known.

CREEPING THISTLE

Description

Creeping Thistle (*Cirsium arvense*) is a perennial weed with typical thistle-like prickly leaves and white creeping underground roots. These roots give rise to new shoots at a distance from the parent. This is probably the only thistle likely to be troublesome in gardens, although the Spear Thistle (*C. vulgare*), a biennial, may occasionally occur in lawns and paddocks. Spear Thistle is easily cut out with a hoe and will not regrow.

Occurrence

Creeping Thistle is one of the worst perennial weeds in this country and indeed in many others. Its rapid vegetative spread by the roots makes it a serious pest in most situations. It grows in a variety of habitats on most soils throughout the British Isles.

Reproduction and Spread

Creeping Thistle can produce seeds, occasionally in abundance, but the seeds are of relative unimportance in gardens because flowering is easily controlled. The main method of reproduction is by the creeping underground root system. In loose soil, the roots can creep horizontally for considerable distances. When broken up, each fragment can give rise to new plants. The creeping roots are very brittle so that even with careful digging many fragments are left behind in the soil.

Control

Control of this weed is not easy. Where cultivation is possible, this should be carried out frequently, but it will not lead to any permanent effect. Dichlobenil is very effective and can be used under certain ornamental and fruit bushes, provided that they are established. When thistle shoots start appearing in rockeries and flower beds, it is perhaps easiest to apply MCPA or 2,4-D mixed with other chemicals (as for lawns) to each group of leaves with a paintbrush. This may need to be repeated. Great care must be taken to avoid touching other plants. On lawns, these herbicides can be sprayed overall; on paths, dichlobenil can be used.

SOW-THISTLES

Description

Spiny Sow-thistle (*Sonchus asper*) and Annual
Sow-thistle (*S. oleraceus*) are two very similar annual or
overwintering weeds. Spiny Sow-thistle has spiny leaves,
which are glossy and green. The leaf base where it joins
the stem is rounded and pressed flat against it. Annual
Sow-thistle, on the other hand, has dull bluish-green
leaves which are not spiny, and the base of the leaf
projects as straight points beyond and on either side of
the stem. The flowers of Annual Sow-thistle are a paler
yellow than those of Spiny Sow-thistle. The stems of both
exude a milky juice when broken.

Occurrence

These weeds are troublesome in cultivated soil, waste
places, and commonly in gardens. They both occur
throughout the British Isles, but are less widespread in
Scotland.

Reproduction and Spread

Seeds of both plants germinate in spring and autumn and
flowering occurs from June to August. Much seed is
formed; up to 6000 seeds per plant have been recorded
for Annual Sow-thistle, and 18 000 for Spiny Sow-thistle.
Their seeds, like those of the Perennial Sow-thistle, are
very light and have a plume of hairs, which assists
dispersal by the wind.

Control

As they occur mainly in vegetable and flower beds and
generally as single plants, hoeing is probably the best
method of control. Hand pulling of odd plants is difficult
as the stems almost invariably break and subsequently
grow again from the base. Both of these weeds are
susceptible to various herbicides, where they can be
used (see under Annual Weeds of Cultivated Land on
page 82 at the end of the book). As Sow-thistles are
rarely abundant, the use of herbicides will probably be
to control other weeds. On paths, control can be by
simazine, dichlobenil or the paraquat/diquat mixture.

FAT HEN

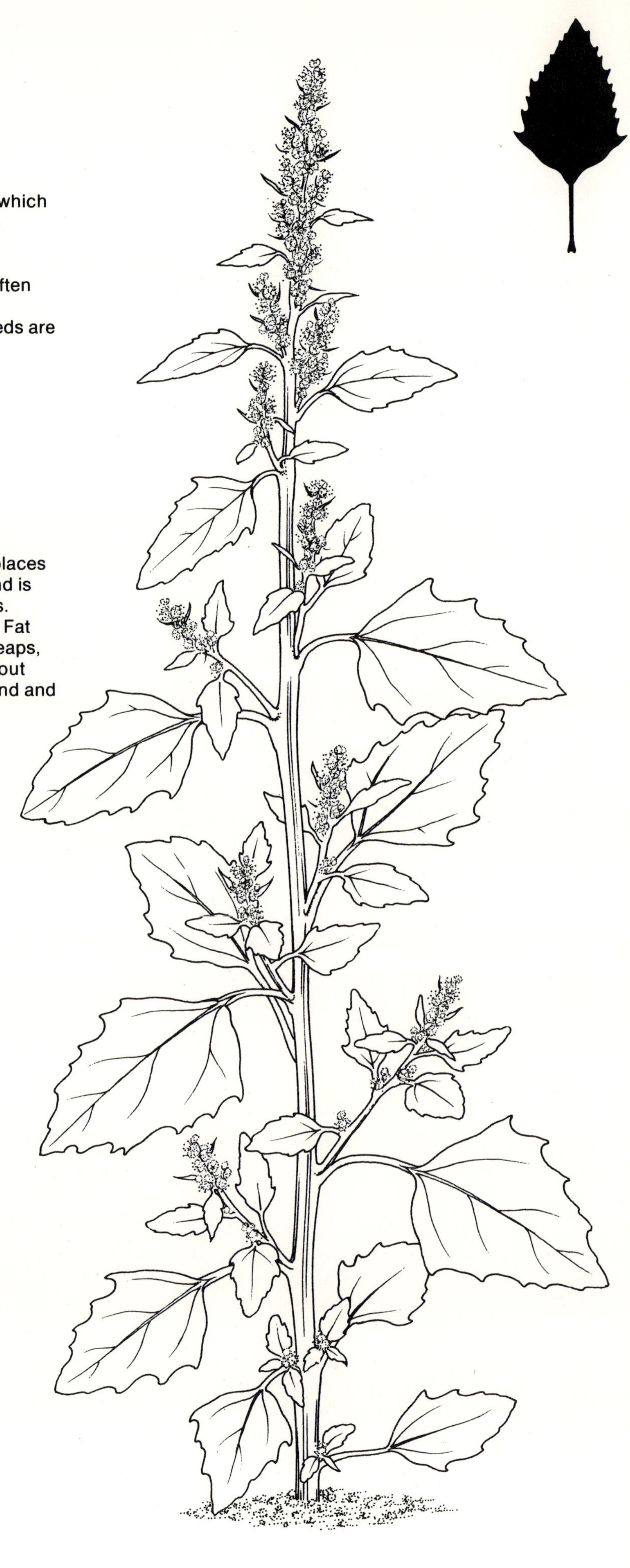

Description

Fat Hen (*Chenopodium album*) is an annual weed which grows very straight and can reach 3 feet (90 cm) in height. The leaves are variable in shape, but are characteristically rather fleshy and silvery-white in appearance, especially when young. The stem is often reddish in colour. The flowers are greenish and inconspicuous although in dense clusters. The seeds are generally black and shiny.

Occurrence

Fat Hen is one of the commonest weeds of waste places and cultivated land, both farm and horticultural, and is very frequently found in gardens and on allotments. Although drought hardy, it is readily killed by frost. Fat Hen is a gross feeder and grows well on manure heaps, although it can grow on all soils. It occurs throughout the British Isles, although less commonly in Scotland and Ireland.

Reproduction and Spread

The seedlings emerge rather late, appearing in late April onwards, with a further flush in August. The plant grows rapidly, especially on fertile soils, and flowers from July to September. Its partiality for manure heaps frequently leads to its seed being spread about on cultivated soil in manure. The average number of seeds produced by each plant is about 3000, but up to 28 000 have been recorded; they are dormant when shed and some are capable of surviving about 30 years in the soil. It is, therefore, essential to prevent seeding. If seed is formed but not yet shed, burning is the only remedy, for seeds can survive in the compost heap.

Control

Fat Hen is easily killed by hoeing when young, especially in dry weather. As the plants grow, the stem toughens and hand pulling is a possibility, provided that there are not too many. For other methods of control, see under Annual Weeds of Cultivated Land (on page 82) and the table (on page 86) at the end of the book.

WHITE DEAD-NETTLE

Description

White Dead-nettle (*Lamium album*) is a hairy perennial weed with creeping underground stems. The leaves are in opposite pairs and are coarsely toothed. The stem is square and the flowers at the top of the shoots are large, white and conspicuous.

Occurrence

White Dead-nettle is frequent in waste places and roadsides and is often a weed in gardens where, because of the underground stems, it can be a persistent problem in flower beds. It is very common in England, but is much less frequent elsewhere in the British Isles.

Reproduction and Spread

The underground stems are thick, white and often branched. Fragments are capable of regrowing new plants and pieces are frequently left behind after digging. Seeds are produced and these also help the weed to persist.

Control

As the underground stems are generally shallow in the soil, they are quite easily forked out, but care must be taken to remove all the pieces. Among desirable plants, it is awkward to remove as it has a tough and extensive root system. Its response to herbicides is not known, but dichlobenil may suppress the shoots amongst certain ornamental and fruit bushes, where it can be used if the bushes are established. If eradication is not possible, at least prevent seed being shed.

ANNUAL NETTLE

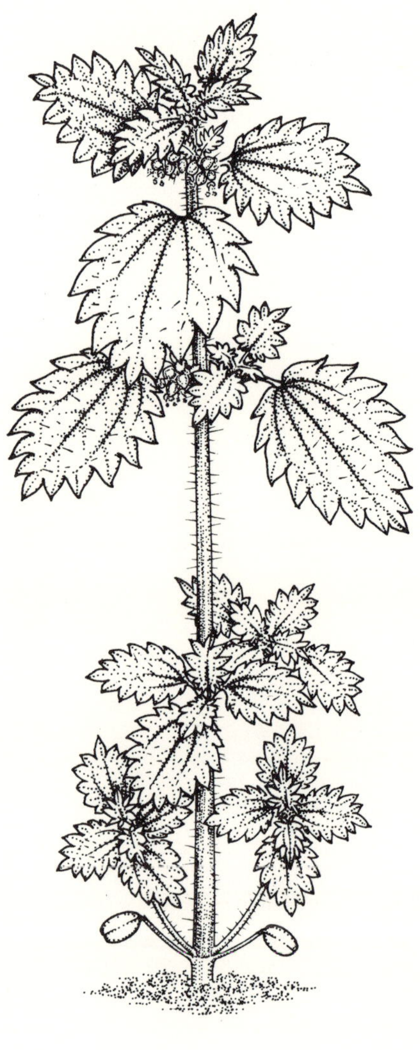

Description

Annual Nettle (*Urtica urens*) is a smaller version of the
Stinging Nettle, but less common. Although they both
have stinging hairs, Annual Nettle differs from Stinging
Nettle in having white roots, not yellow, and the stems
never creep along the ground. It grows to a maximum
of 2 feet (60 cm) tall. The flowers are green and not very
conspicuous.

Occurrence

When Annual Nettle does occur in gardens it can be a
serious weed because of its abundance. It occurs on light
soils in cultivated ground, mainly in gardens and
horticultural nurseries. It is widespread in the British Isles
and is especially common in the east.

Reproduction and spread

Germination of seed occurs in spring and autumn and
flowers are quickly produced and open between June
and September. Over 1000 seeds can be formed by each
plant and it can rapidly build up into a dense infestation,
partly because it has a short life-span and partly because
it begins to produce ripe seed at a very early stage (at
about six leaves).

Control

There is only one cardinal rule for controlling this weed
and that is to prevent seed production completely, for
the seeds can survive for several years in the soil. As
seeding can occur at the six-leaf stage, constant hoeing
is essential. Hand pulling (with gloves of course) should
be carried out when hoeing is difficult and each plant
must be removed at an early stage. Seedling establishment
can be prevented by chlorpropham, 2,4-DES,
chloroxuron and propachlor, where these can be used;
see under Annual Weeds of Cultivated Land on page 82
at the end of the book and on page 86 of the table of
weed susceptibility.

STINGING NETTLE

Description

The Stinging Nettle (*Urtica dioica*) is a coarse perennial weed, which must be familiar to everyone because of its irritating stinging hairs. The roots are stout, spreading and bright yellow. The stems, which are purplish at first and creep along the soil surface, become erect and green in spring. Annual Nettle, which also has stinging hairs, has a white root and the aerial stems never creep horizontally. Stinging Nettle is much larger than Annual Nettle and can grow to 4 feet (120 cm) tall. The flowers of both plants are green and inconspicuous.

Occurrence

Stinging Nettle grows on all soils, especially rich ones, and in many situations, such as rough grass and waste places, hedges and neglected areas and around derelict buildings. In gardens, it occurs mainly in orchards, paddocks, hedges and shrubberies. It is abundant, and generally distributed throughout the British Isles.

Reproduction and Spread

The creeping stems increase the size of the clumps slowly but surely. They become erect in spring and give rise to the flowering shoots. Flowering takes place from June to September, and seeds can be produced in great numbers. The seeds can survive for several years in the soil, so seeding should be prevented where possible.

Control

For large patches of nettles, control can be achieved by cultivation or by herbicides. Odd plants are, however, most easily removed by forking out, using the fork as a lever, or by undercutting with a spade. MCPA or 2,4-D mixed with other chemicals, or 2,4,5-T alone or with 2,4-D can be sprayed on to big areas in paddocks. This may need to be repeated. Care must be taken to avoid drifting spray onto nearby desirable plants. Dichlobenil can be used in certain ornamental and bush fruits if well established or on paths and drives. Sodium chlorate, which gives total plant kill, can be used in waste areas where there are no ornamentals, vegetables or trees.

COLTSFOOT

Description

Coltsfoot (*Tussilago farfara*) is a perennial weed with large angular leaves, which are covered with a whitish web of hairs when young, but become green later on. The underground stems are thick, white and extensively creeping. The flowers, which are borne on short stems, appear early in the year before the leaves. They are bright yellow and resemble dandelions.

Occurrence

Coltsfoot is a widespread weed of agricultural land and can be troublesome. It can also be an intolerable nuisance in gardens, because of its underground stems. It grows on most soils, and is found throughout the British Isles.

Reproduction and Spread

When plants are established, the aerial stems emerge in March and the flowers open in March and April. These are followed closely by the leaves, which in dense infestations can cover the ground completely. The number of seeds produced can be very large. The seeds are wind-dispersed and are capable of travelling great distances. There is no dormancy and germination is immediate if the conditions are suitable. The seeds are very short-lived and are only capable of germinating for three to four months.

Control

The short life of the seed does permit complete eradication of the weed, provided that the creeping underground stems can be dug out completely. However, as these can penetrate for several feet in to the soil, eradication is very difficult. The creeping stems are thick and fleshy, and small fragments can give rise to new plants. There is also the risk of fresh seed blowing in again from a distance. The weed is not very susceptible to weedkillers. Sodium chlorate and 2,4,5-T will have some effect, but care must be taken over their use; see Chapter 9 on herbicides (on page 89) at the end of the book. Dichlobenil can be used among certain ornamental and fruit bushes, if they are well established, and also on paths and drives.

BLACK NIGHTSHADE

Description

Black Nightshade (*Solanum nigrum*) is a bushy annual weed growing to about 2 feet (60 cm) tall. It has broad leaves, which are usually hairy. The flowers are white with yellow centres and are borne in pendant clusters. The fruits, about the shape and size of a pea, are green at first but turn black and juicy when ripe. The plant is somewhat poisonous.

Occurrence

Black Nightshade grows in waste places and cultivated ground in farms and gardens, for it likes rich soils. In distribution it is confined to England and Wales, being most common and widespread in the south and east of England.

Reproduction and Spread

Compared to most weeds, the seedlings emerge late, the majority appearing in mid-summer from May to July. No germination occurs in the autumn, for it is sensitive to frost. The plant flowers from July to October. It seeds prolifically, each plant producing on average about 10 000 seeds. The seeds can survive up to about 40 years burial.

Control

As Black Nightshade is a plant of cultivated soil and withers quickly in warm weather, hoeing is probably the best method of control. Hand pulling of small numbers is feasible. It is resistant to many herbicides, such as 2,4-D, MCPA, mecoprop, dichlorprop and dichlobenil, but it can be controlled by simazine, chloroxuron and the paraquat/diquat mixture, where these can be used. See under Annual Weeds in Cultivated Land on page 82 at the end of the book, and on page 86 of the table of weed susceptibility.

GALLANT SOLDIER

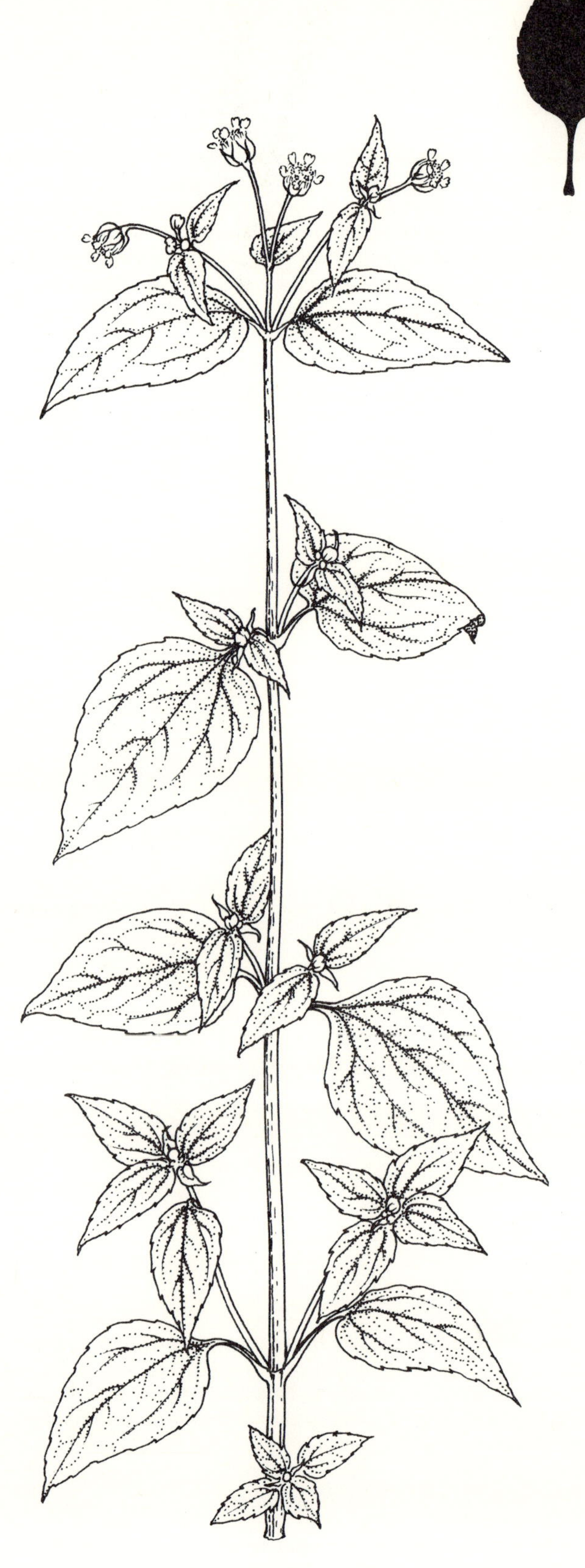

Description

Gallant Soldier (*Galinsoga parviflora*) is an erect annual weed, which can grow to 3 feet (90 cm) tall. The flowers are clustered together in heads, rather like daisies, but there are usually only five white petals surrounding the yellow centre. Another very similar weed, Shaggy Soldier (*G. ciliata*), differs in having the stem densely covered in long hairs.

Occurrence

Gallant Soldier was introduced from South America and was first recorded as a garden weed at Richmond in 1860. Since then, it has spread slowly outwards and is now widespread in Kent, Essex, Surrey, Middlesex, Hertfordshire and Bedfordshire, and occurs sporadically elsewhere. It grows in gardens and horticultural nurseries where it is occasionally a serious weed. Seeds in the soil around the roots of transplanted ornamentals probably account for most of its spread, for it hardly occurs outside gardens and nurseries, although its particular association with gardens may be due in part to its liking for rich soils.

Reproduction and Spread

Gallant Soldier is an annual weed and so relies on seeds alone for reproduction and spread. The plant flowers from May to August. A single plant can produce up to 15 000 seeds, although 2000 is probably nearer the average. As there can be more than one generation in a year, it can increase in numbers very rapidly. The seeds have no dormancy and most germinate soon after shedding. However, under some conditions they can remain alive in the soil for up to ten years.

Control

The plant can readily be controlled by hand pulling or hoeing, which should be carried out before any seed is formed. The plant is susceptible to 2,4-DES, propachlor and chloroxuron, where they can be used; see under Annual Weeds of Cultivated Land on page 82 at the end of the book. Simazine can be used in shrubs and bushes.

ROSEBAY WILLOW-HERB

Description

Rosebay Willow-herb (*Chamaenerion angustifolium*) is a tall perennial weed growing to about 4 feet (120 cm) in height. It has creeping roots, which can produce new shoots anywhere along their length. The aerial stems have numerous leaves, which are long, narrow and slightly toothed. The flowers are bright purple. The seed capsule is long and narrow, and splits open when ripe to allow the wind-borne seeds to escape.

Occurrence

Rosebay Willow-herb is common throughout the British Isles, excepting Ireland, and grows in a wide range of situations – especially in gardens, old woodland, waste places and around derelict buildings.

Reproduction and Spread

The creeping underground roots spread the plant locally and so build up dense clumps. The seed production is enormous and as the seed is wind-borne it is spread about very widely. It is particularly associated with bare soil.

Control

Hoeing is not a satisfactory method of control, but as the creeping roots do not go too deep in the soil the plant can be forked out without too much trouble. However, when growing among ornamentals it is a persistent problem, which is very hard to eradicate. Dichlobenil can be used around certain ornamental and fruit bushes, provided that these are established. In rough neglected areas, sodium chlorate, which gives total plant kill, is very effective against Rosebay Willow-herb.

BROAD-LEAVED WILLOW-HERB

Description

Broad-leaved Willow-herb (*Epilobium montanum*) is a
perennial weed, which grows to a maximum of 2 feet
(60 cm) in height and has small pink flowers. The seed
capsule is distinctive, as it is very long and thin and, when
ripe, splits lengthways into four segments. The seeds,
each with a long white plume of hairs as they are
wind-dispersed, are very noticeable. There are several
Willow-herbs in this country, all very similar. This is the
one that occurs most frequently in gardens. It is not a
weed of great importance, but it is unsightly, especially
when seeding.

Occurrence

Although possibly more of a woodland plant, it grows in
a wide range of situations, including the top of dry stone
walls. It can occur almost anywhere in a garden. It grows
on all soils throughout the British Isles.

Reproduction and Spread

The seeds, which can be blown great distances by the
wind, germinate both in spring and autumn. In the
autumn, established plants produce small rosettes of
leaves at the end of short creeping stems. These
overwinter and form new flowering stems in the spring.
The plants therefore spread vegetatively as well as by
seed. Flowering occurs from June to August and the
seeds are shed over many weeks.

Control

The most effective method of control is to dig up each
plant separately, making sure than none of the short
creeping stems formed in the autumn remain below
ground, for they are brittle. Hoeing too and even hand
pulling can be useful, although regrowth can occur from
the base of the stem. The herbicides 2,4-D and MCPA
have virtually no effect; the paraquat/diquat mixture is
effective on the rosettes of leaves, and dichlobenil, where
it can be used, such as on paths, is also effective.

BLUE SOW-THISTLE

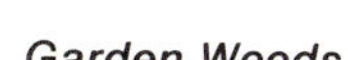

Description

Blue Sow-thistle (*Cicerbita macrophylla*) is a perennial weed with flowering shoots growing to about 5 feet (150 cm) tall. The lowest leaves, which are a bluish green colour, consist of a large terminal lobe and two smaller ones on either side of the winged leaf stalk. The flower heads, which are up to 1½ inches (4 cm) across, are a most attractive light blue. Apart from flower colour, it is similar to Perennial Sow-thistle.

Occurrence

Introduced from the Caucasus, probably in the sixteenth or seventeenth century as an ornamental, it has become firmly and widely established in a number of counties. Although very local in occurrence, it is a great nuisance where it does occur, especially in small gardens.

Reproduction and Spread

Establishment and spread are entirely from creeping underground stems, as seed does not appear to be produced; this is fortunate, as the plant would otherwise be a much greater and more widespread problem. When established, it forms a thick mass of roots among which creep the underground stems, which are whitish in colour. These stems creep quite rapidly and spread the weed about. Long-distance dispersal is lacking, apart from deliberate or accidental transplanting by gardeners.

Control

The plant can be controlled by digging, but it is essential to remove all fragments of the underground stems, because it will otherwise quickly re-establish itself. If digging is carried out thoroughly, the plant can be eradicated, for no seed is produced. When growing among ornamentals the position is difficult, for hoeing probably has little effect. Its response to weedkillers is unknown, but painting the leaves with MCPA or 2,4-D may help, judging by their effects on closely-related plants. If eradication is impossible, its area should be restricted by putting a continuous barrier around it in the soil, otherwise it is liable to take over the whole garden.

ANNUAL MERCURY

Description

Annual Mercury (*Mercurialis annua*) is an annual weed growing to a maximum of 2 feet (60 cm) tall. The leaves are in opposite pairs on the stem and the flowers are green and inconspicuous. Male and female flowers are normally on separate plants. The plant is poisonous and drying the plant does not affect the toxins, so fresh or dried, it should not be given to pets or farm animals.

Occurrence

Annual Mercury occurs only sporadically in the British Isles and mainly in the south of England and in East Anglia. It prefers light sandy soils, growing in waste places, fields and especially in gardens, where it can be very troublesome.

Reproduction and Spread

The seeds germinate in late spring and early summer. Flowering is from July to October, and the seeds ripen from August onwards. When ripe, the seed pods break open suddenly and shoot the seeds out forcibly. The spread of the plant can therefore be quite rapid, especially as more than one generation can occur in a single year. Seeds are set in abundance and they germinate readily.

Control

Control by hand pulling or hoeing is easy, but they must be carried out before any seeds are set. At later stages the plants should be burnt, even if the seed appears unripe, for some of them can germinate. It is essential that seed shed be prevented. The response of the weed to most herbicides is unknown, but dichlobenil and simazine, where they can be used, are effective; see page 89 at the end of the book.

SLENDER SPEEDWELL

Description

Slender Speedwell (*Veronica filiformis*) is a perennial
weed with numerous surface-creeping stems. It differs
from all other speedwells in that the leaves are
kidney-shaped. The flowers are bright blue, indeed they
are very attractive, unless the plant is growing as a weed
in your garden.

Occurrence

Slender Speedwell occurs in lawns, on drives, paths and
in flower and vegetable beds, for it creeps indefatigably.
It was introduced from Asia Minor about 1800, but was
not widely planted in rockeries and flower beds until
about 1920, when its true nature became only too evident
and gardeners ever since have been trying to get rid of
it. It is more or less confined to gardens, although it
occasionally escapes into hedges and roadsides
adjacent to them. It is scattered but widespread in the
British Isles, being commonest in southern England.

Reproduction and Spread

Fortunately the plant is self-sterile and, as most of the
plants in this country appear to have originated from a
single strain, seed is not formed at all. Consequently, all
dispersal is vegetative.

Control

The fact that seed is not produced gives the advantage
that, if the weed is successfully eradicated, the plant will
not return as seedlings. In cultivated soil, it can be
destroyed by hoeing or digging. In lawns, it has proved
to be a difficult plant to kill, but a recently-developed
mixture of ioxynil and mecoprop now gives good control.
Lawn sand and tar oil can also be used; see page 89
at the end of the book.

SPEEDWELLS

Description

Wall Speedwell (*Veronica arvensis*) and Field Speedwell (*V. persica*) are two small, closely-related, annual weeds of gardens. Wall Speedwell has erect flowering spikes with small flowers of deep blue, while Field Speedwell has outward-spreading stems with the flowers borne singly in the angles of the leaves. Field Speedwell is larger than Wall Speedwell in leaf size, stem length and flower size, and the flowers are bright blue with one petal a paler blue or even white.

Occurrence

Both are frequent in gardens, but Field Speedwell is possibly the greater nuisance. They both grow on all soils, especially cultivated ones, although Wall Speedwell can also occupy many other habitats as the name suggests. Both occur widely throughout the British Isles, which is surprising for Field Speedwell, because it was first recorded in this country only in 1825. It has since then spread rapidly and is now one of the commonest speedwells.

Reproduction and Spread

Wall Speedwell flowers from March to October. It produces a considerable number of seeds, which can survive for several years in the soil. Field Speedwell can flower the whole year round and may have two generations in the course of a single year. It produces on average about 2000 seeds per plant and these can survive for five to ten years in the soil. Seedlings appear mainly in spring and autumn, but can be found in any month.

Control

Control by hoeing is easy as they are shallow-rooted weeds. Both are resistant to many herbicides, but the paraquat/diquat mixture can be used on them successfully when the plants are young. Propachlor and chlorpropham, where they can be used, will prevent seedling establishment; see page 82 at the end of the book.

DAISY

Description

The Daisy (*Bellis perennis*) is a small weed which in gardens is found only in lawns. The plant forms clusters of shiny, blunt-ended, dark green leaves. The well-known and conspicuous flower heads, with yellow centres and white petals around the edge, open during the day and close at night, hence the name 'Day's Eye'.

Occurrence

The Daisy is one of the commonest plants and grows in short grass, especially in close-mown lawns where its small size is no disadvantage. It can withstand trampling and treading so is frequently found along the edges of tracks and paths. It occurs on all soils and is locally abundant throughout the British Isles.

Reproduction and Spread

The plants give rise to prostrate short-creeping shoots which enlarge the colony gradually. Seed is produced whenever the plants are allowed to flower. Flowering occurs between March and October and a hundred or more seeds are produced by each flower head. The seeds are dispersed by wind, in mud on boots and by birds. The seed is very viable and germinates both in spring and autumn.

Control

Although it seems a pity to destroy such a harmless plant, which gives such pleasure to children, it can disfigure an otherwise perfect lawn. For those who wish to remove it, it is easy to control, because it is susceptible to most lawn weedkillers.

DOCKS

Description

Curled Dock (*Rumex crispus*) is a large perennial weed growing to about 3 feet (90 cm) tall. The root is thick, yellow-brown in colour and tough. The dark green leaves are large and usually curled or undulating at the margins. The Broad-leaved Dock (*R. obtusifolius*), has a similar tap root, but the leaves are much broader, have flat margins and rounded lobes at the base.

Occurrence

Curled Dock grows equally well in cultivated land and grassy places. It occupies many other habitats, such as waste places and roadsides. In gardens, it is most likely to occur in rough grassland, such as paddocks and orchards. It grows throughout the British Isles on all soils except the most acid.

Reproduction and Spread

Established plants start growth anew from their roots in February and March and flowering occurs between June and October. Unlike the Broad-leaved Dock, the Curled Dock often dies after flowering. However, it is best not to let it flower as seed production in undisturbed places can reach 30 000 per plant per year. Moreover, the seeds are capable of prolonged survival, anything up to 80 years in the soil. Seedlings can emerge in various months, but germination occurs mainly in spring and late summer.

Control

To prevent seeding, flowering shoots should be cut down before the flowers open, but this will need to be repeated more than once during the year. The best way to kill large plants is to dig out the top 4-6 inches (10-15 cm) of the tap root, which is the regenerative part. Shoots are not produced from deeper than 6 inches (15 cm). Hand pulling is impracticable as the roots are too tough to pull out. In rough grass, various of the lawn weedkillers can be used, but may need to be repeated. Dichlobenil can be used in shrubberies.

RED DEAD-NETTLE

Description

Red Dead-nettle (*Lamium purpureum*) is a small annual weed growing to about 18 inches (45 cm) tall. The whole plant is frequently tinged with purple. The branches grow at first along the ground and then turn upwards. The flowers, which open between March and October, are purple in colour and occur in clusters at the top of the stem. The leaves are shaped like a shield or heart and the lower ones do not have such a long stalk as those of Henbit Dead-nettle (*L. amplexicaule*), which also differs in having leaves which are more or less circular in outline. The two plants grow in similar situations.

Occurrence

Although occurring generally in cultivated land, Red Dead-nettle is less common in farmland than in gardens and horticultural holdings. It grows on all soils and is found throughout the British Isles.

Reproduction and Spread

Seed production is not as great as that of many other weeds and this may be the reason why it is rarely abundant. The seed is thought to survive for some years in the soil and so, as with all weeds, seed production should be prevented if possible. Germination can occur throughout the year, although it is mainly in spring and autumn. The plant is very hardy and can survive severe frosts.

Control

The plant can be pulled up by hand, but the stems may break, so hoeing is probably the best method of control. Red Dead-nettle, like Henbit Dead-nettle, is resistant to many herbicides that can be used in gardens, but is susceptible to simazine and propachlor, where these can be used. See under Annual Weeds in Cultivated Land on page 82 at the end of the book.

HENBIT DEAD-NETTLE

Description

Henbit Dead-nettle (*Lamium amplexicaule*) is a small
annual weed growing to about 10 inches (25 cm) tall.
The lower leaves are very long-stalked, hairy, and roughly
circular in shape. The leaf-like bracts, which bear the
flowers in their axils, are stalkless and clasp the stem.
The flowers, which are grouped in small clusters, are
rosy-purple and the colour is deepest and most apparent
just before the flowers open. Red Dead-nettle grows in
similar places, but differs in having leaves shaped more
like a shield and shorter stalks. The whole plant is often
tinged with purple.

Occurrence

Henbit Dead-nettle is a local weed of cultivated land,
especially horticultural nurseries and gardens, where it
is occasionally a nuisance. It occurs on all soils, but
prefers light sandy ones, and is most frequent in southern,
central and eastern England and eastern Scotland.

Reproduction and Spread

The flowers open between April and August and the
seeds are ripened in succession. Freshly-shed seeds
are dormant. There is therefore every reason to prevent
seed being formed, because the dormancy could lead to
a rapid build up in the number of seeds in the soil,
especially as the plant is capable of up to three
generations in a single year, under good conditions.

Control

Hand pulling is likely to break the stems rather than
uproot the plants and, furthermore, it is very tedious
when there are many. Hoeing is therefore the best
method of control. It is said that stem fragments can
re-root to form new plants, so they should not be left on
the soil surface, but raked up and composted. The plant
is resistant to many herbicides, but simazine and
dichlobenil are effective on paths and under certain
ornamental and fruit bushes, which must be established.

HOARY CRESS

Description

Hoary Cress (*Cardaria draba*) is a perennial weed
growing up to a maximum height of about 3 feet (90 cm).
It has a system of thin creeping roots, which give rise
to new aerial shoots at a distance from the parent plant.
The leaves are variable in size and shape, but are
generally grey-green and oval. The flowers are white.

Occurrence

Hoary Cress grows in farmland and waste places,
especially roadsides and can also occasionally occur in
gardens. It was introduced into Kent from Europe in 1809
and since then has been spreading steadily. It is now
common in much of East Anglia and south-eastern
England; it occurs occasionally up to a line from the
Humber to the Severn, but north and west of this it is
only very scattered.

Reproduction and Spread

Hoary Cress produces seeds, but relies much more on
the creeping roots, which produce new shoots around
the parent plant. As a result it tends to form large
patches. The creeping roots can penetrate about 2 feet
deep into the soil.

Control

It is largely a waste of time to try and dig out this weed
and hoeing can never be more than a temporary control
measure. As it is quite susceptible to both MCPA and
2,4-D, these should be used wherever possible; it may
mean painting one of these herbicides (as mixed with
other chemicals for lawns) on to the leaves with a
paintbrush, as this weed usually grows in among
ornamentals or in vegetable beds. Care must be taken
of course to avoid other plants. Dichlobenil and sodium
chlorate are useful where they can be used; see Chapter
9 on page 89 at the end of the book.

WINTER HELIOTROPE

Description

Winter Heliotrope (*Petasites fragrans*) is a perennial plant with very thick creeping underground stems. Apart from the flowering stems, all the stems of this plant remain below ground, with only the leaves emerging. The flowers, which are light mauve and vanilla-scented, are grouped in heads. The leaves, which are about 6 inches (15 cm) across, resemble those of Coltsfoot (*Tussilago farfara*), but are more rounded in shape and persist throughout the winter, while Coltsfoot leaves die back in October.

Occurrence

Originating in the western Mediterranean, Winter Heliotrope was introduced to Britain in about 1800 as a winter-flowering ornamental for gardens. It has since then been much planted and is now widely distributed in the British Isles, although commonest in the south. It is still planted occasionally both as an ornamental and for the leaves, which are sometimes used in the flower trade. Winter Heliotrope can be a serious nuisance in gardens, for once it starts spreading it will invade almost anywhere, such as lawns, paths, rockeries and tennis courts. It also occurs as a naturalised escape in large patches on roadsides and waste ground, where it is very obvious, but of little importance.

Reproduction and Spread

Although widely planted, it is by no means always a pest. In many gardens it hardly spreads at all, but in others, particularly in the south, it can spread very rapidly – up to a maximum of 3 feet (90 cm) in a year. Seed production is fortunately rare as the sexes are on separate plants. The seed, when it is produced, is wind-borne.

Control

Control of this weed is difficult in gardens because it grows in places one doesn't wish to disturb. However, digging is one of the best ways of controlling it, for the stems are large and fairly close to the soil surface. Its response to herbicides is unknown, but a close relative is resistant to many.

CREEPING BELLFLOWER

Description

Creeping Bellflower (*Campanula rapunculoides*) is a
perennial weed with large spires of attractive purple
flowers, which are similar to those of other species of
Campanula. It has large fleshy roots like carrots, but they
are white and can regrow rapidly if the shoots are cut
down.

Occurrence

The plant grows scattered here and there throughout the
British Isles in fields, grassy places and in gardens,
particularly old-established ones where it may originally
have been planted. It appears to prefer heavy soils.
Introduced in earlier times as an ornamenal plant and
widely grown, it is now well established in the country,
although no longer in fashion because of its tendency
to become a serious weed.

Reproduction and Spread

It sets seed if allowed to flower and although germination
is said to be poor, sufficient seed is normally set to allow
a good crop of seedlings in the following and later years.
As the roots produce side shoots, this gradually
increases the size of the patch, and as the seeds are not
actively dispersed, the clumps tend to be dense so that
nothing much else can grow there. The leaves appear
above ground in February and March and erect flowering
shoots are produced later on. The flowering stems grow
to about 3 feet (90 cm) in height. The flowers open in
June and flowering continues until September, or even
later in a warm autumn. After flowering is over, a new
crop of small shoots appears above ground, but these
die back again in early winter

Control

Control is difficult and can only be achieved by digging
up each plant. As seed can remain dormant in the soil
for several years and the roots regenerate readily, this
is inevitably a prolonged process. Its response to
herbicides is not known.

CANADIAN FLEABANE

Description

Canadian Fleabane (*Conyza canadensis*) is an annual
weed which grows up to 3 feet (1 metre) tall. The stems
are rigidly erect and bear a large number of long, narrow
leaves. The flowers are small and clustered into heads
which are a pale yellowish white in colour. The seeds
have a parachute of hairs which becomes visible as the
seeds ripen, and the flower heads then appear fluffy.

Occurrence

Canadian Fleabane is a weed of gardens and waste
places, especially in cities where it can grow in odd
corners and even between paving stones. An
introduction from North America, it first appeared in
London about 1700, but it has only been spreading more
widely since about 1900. It is now frequent to the south
and east of a line from the Wash to the Isle of Wight.
Beyond the line, the weed becomes more scattered, but
it is still spreading steadily towards the north and west.

Reproduction and Spread

Canadian Fleabane flowers in summer from August to
September. It produces a large number of seeds which,
with their parachute of hairs, are very widely spread by
the wind. Seedlings appear in the autumn and overwinter
as rosettes of long narrow leaves.

Control

Although a weed eminently suited to colonising
neglected gardens, it is very vulnerable to ordinary
garden maintenance, for it has a long period of
development when it is easy to hoe out. Also, when
flowering starts, the erect flowering stem just lends itself
to hand pulling. It is susceptible to 2,4-D and MCPA,
where these can be used, but its response to other
herbicides is unknown. Simazine can be used on paths
and in shrubberies.

PLANTAINS

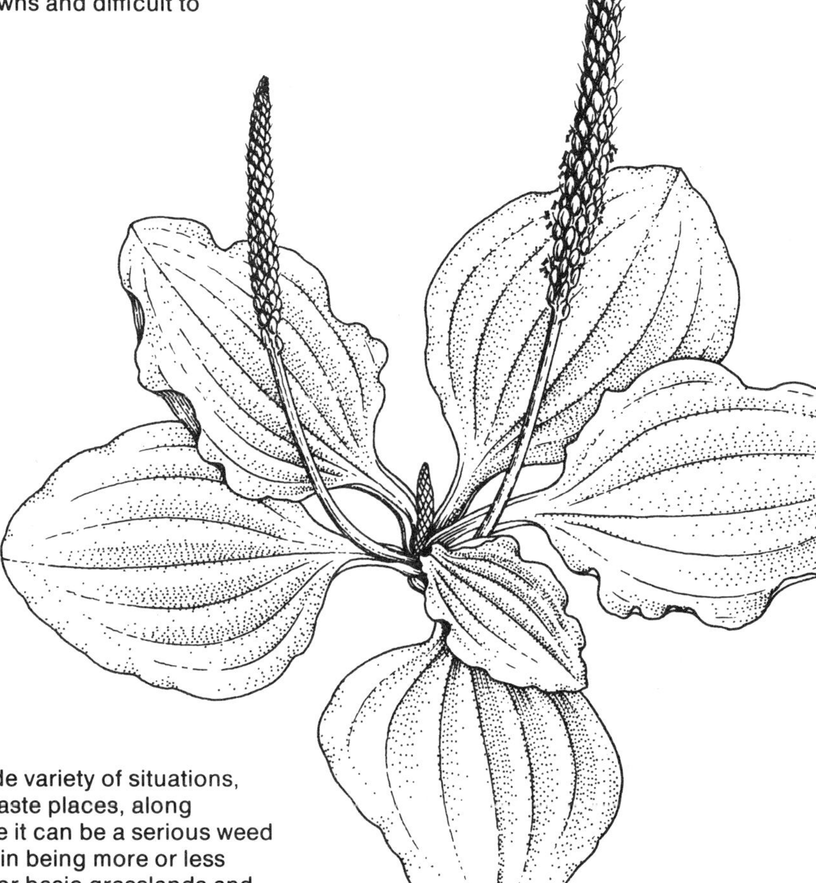

Description

Greater Plantain (*Plantago major*) and Hoary Plantain (*P. media*) are two closely-related perennial plants, which can occur as weeds of grassland or lawns. They both have broad leaves and form leafy rosettes. Leaves of Greater Plantain have stalks, which can be as long as the blade; while leaves of Hoary Plantain, which are grey green and hairy, are virtually stalkless. They both have clusters of small flowers borne on short unbranched stalks, which are unsightly in lawns and difficult to remove by mowing.

Occurrence

Greater Plantain occurs in a wide variety of situations, such as on cultivated land, in waste places, along pathways, and in gardens where it can be a serious weed of lawns. Hoary Plantain differs in being more or less confined to close turf in chalky or basic grasslands and never occurring in cultivated ground, but it is sometimes found along paths as it can withstand trampling. These two broad-leaved Plantains are particularly suited to lawns because their flat rosettes of leaves can escape mowing. Greater Plantain occurs throughout the British Isles, while Hoary Plantain is more or less confined to England.

Reproduction and Spread

Growth of these Plantains starts in spring and flowering occurs in early summer and continues until autumn, from May to September. Seed production varies with the situation, but average figures have been given for the Greater Plantain as 14 000 per plant per year and for Hoary Plantain as 1000. The seeds of Greater Plantain are thought to survive for several decades in the soil, but those of Hoary Plantain are not so long-lived.

Control

Cutting off the top of the rosettes often causes a ring of new side shoots to develop, so even close mowing can be no cure. It is therefore fortunate that both plants are susceptible to 2,4-D, MCPA, dichlorprop and mecoprop, which are used in various combinations in lawn weedkillers.

ENCHANTER'S NIGHTSHADE

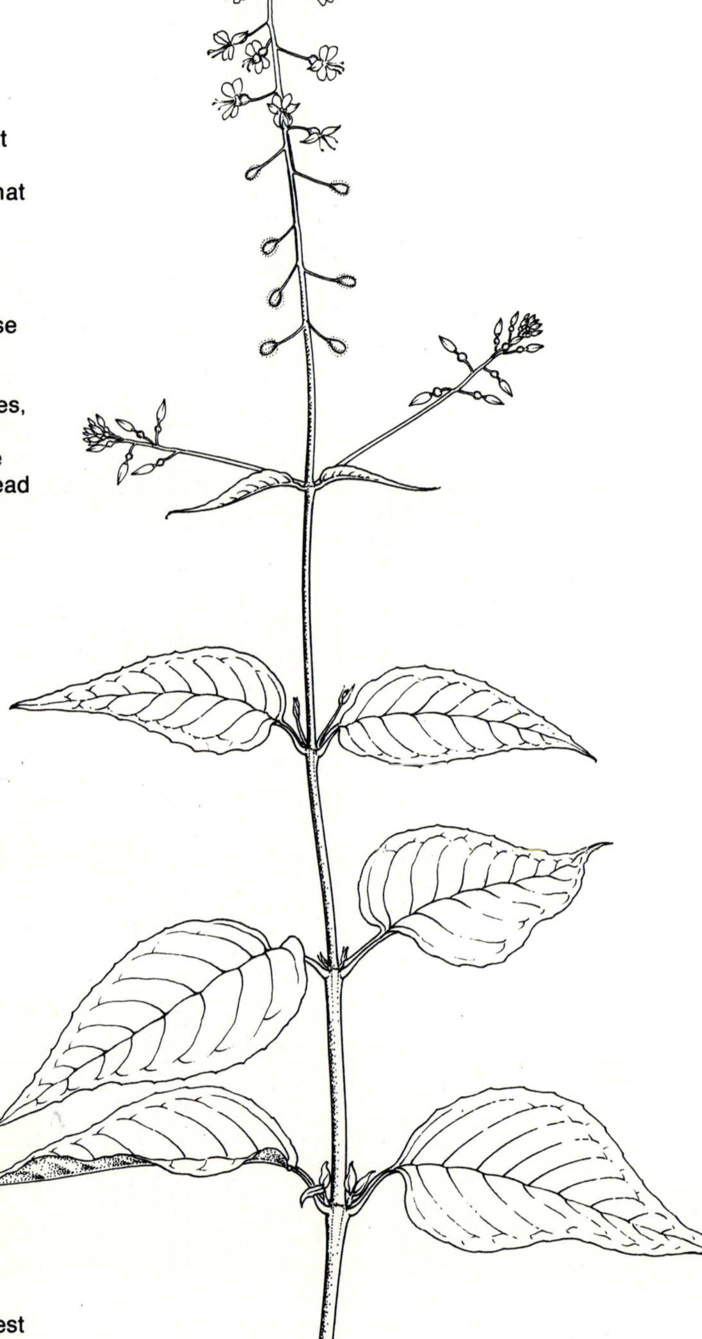

Description

Enchanter's Nightshade (*Circaea lutetiana*) is a perennial plant growing to 2 feet (60 cm) in height. The leaves are a shiny green in young rosettes and may be tinged with red, later they are a dull dark green. The small white flowers with their two projecting stamens are borne in spires at the top of the stems.

Occurrence

Enchanter's Nightshade is normally a woodland plant and appears to prefer shady conditions, but is occasionally a weed of gardens, where it can be a great nuisance among strawberries and herbaceous perennials. It grows on alkaline soils, especially ones that are moist and shady, throughout the British Isles.

Reproduction and Spread

The shoots emerge above ground in spring and give rise to small rosettes of leaves from which grow up the flowering stems. Flowering takes place from June to August. The seeds, which are covered in hooked bristles, frequently become attached to clothing or to the fur of animals so are readily dispersed about the garden. The plant also has creeping underground stems which spread the plant locally. These are soft and white, or occasionally tinged with red. They are very brittle, and when broken up each piece can produce a new plant.

Control

Careful digging out of the underground stems is the best way of getting rid of this plant; but as the underground stems are so brittle, it requires great care to get them out completely. In heavy soils, this may be just about impossible, especially when it is growing amongst other plants. Its susceptibility to weedkillers is unknown.

SUN SPURGE

Description

Sun Spurge (*Euphorbia helioscopia*) is a small golden-green annual weed. The stem is usually unbranched except at the top and rarely grows taller than 1 foot (30 cm) in height. A milky juice gradually oozes out when the stem is broken. The juice is somewhat poisonous. There are several different Spurges in Britain, but the Sun Spurge is one of the two that occur commonly in gardens. The other Spurge which is very common in gardens is the Petty Spurge (*E. peplus*), which is green in colour and not golden green and the leaves are not serrated at the tip like those of the Sun Spurge.

Occurrence

Sun Spurge, although generally insignificant in size, is a noticeable colour and so cannot be ignored when weeding. It doesn't often occur in quantity. It occurs on all soils in cultivated ground on farms and in gardens, and on paths and in waste places throughout the British Isles.

Reproduction and Spread

Sun Spurge is a short-lived annual, which germinates in spring, flowers from May to October and dies in the autumn. The capsules split open abruptly when ripe and fling out the seeds violently, which helps disperse the weed.

Control

Hoeing is one of the best methods of control. Hand pulling usually results in the stem breaking, although the root system is not difficult to pull out, and it can regrow from the base if the stem is broken in this way. Little is known about its susceptibility to herbicides, probably because they have rarely been needed for this weed. It appears to be resistant to most of those tried, but dichlobenil and simazine can be used on paths and among established shrubs.

FIELD BINDWEED

Description

Field Bindweed (*Convolvulus arvensis*) is a perennial weed with climbing stems up to 3 feet (90 cm) long, which twist around other plants. It has an extensive, creeping, underground root system which gives rise to new aerial shoots at a distance from the parent. The bell-shaped flowers are borne in groups of one to three in the angles of the leaves along the stems and can be white, pink or a mixture. It differs from the Hedge Bindweed (*Calystegia sepium*) in that the leaves and flowers are smaller.

Occurrence

Field Bindweed is a very persistent and awkward weed, for it is capable of rapid growth both above and below ground. It grows on all soils in farms, horticultural land, gardens, roadsides and waste places. It does particularly well in dry warm conditions, which also encourage seed production. Field Bindweed is widespread and common in England and Wales, but is much less so in Scotland and Ireland.

Reproduction and Spread

Flowering occurs between July and September and in warm dry years the seeds are formed in quantity. Because of their hard coats, seeds can survive for up to 30 years or possibly longer in the soil. They germinate in both spring and autumn. It is the creeping roots that do most to spread the plant. If undisturbed, they spread further and further, giving rise to new aerial shoots at intervals. If frequently hoed or dug over, the creeping roots are not formed and the plant does not spread. Root fragments do not regrow so readily as those of other weeds, but should be removed.

Control

This weed is very difficult to kill by cultivation, for the vertical roots can penetrate to 20 feet (6 metres) or more in depth. The plant is susceptible to MCPA and 2,4-D, which may have to be applied by paintbrush if the weed is growng among ornamentals. This treatment may need to be repeated. In the vegetable garden, persistent hoeing and annual digging will keep it in check, but will not eliminate it. Dichlobenil is useful among established shrubs. It is also susceptible to oxadiazon; see page 89 at the end of the book.

HEDGE BINDWEED

Description

Hedge Bindweed (*Calystegia sepium*) has very long
climbing stems, which twist around other plants for
support and can reach the tops of hedges. The leaves
vary in shape, but are generally like a broad arrow-head
in outline. The pink, or more usually white, bell-shaped
flowers are about 2 inches (5 cm) across, and borne in
the angles of leaves all along the stems. This Bindweed
is larger and more vigorous than the similar Field
Bindweed.

Occurrence

Hedge Bindweed grows in hedges and in garden
shrubberies, also in nurseries and bush fruit plantations
and along fence lines, from all of which it can be very
difficult to eradicate. It grows on all soils and is
widespread in the British Isles, although commonest in
the south.

Reproduction and Spread

The stems emerge afresh each year during March, and
die back again in late autumn. Flowering occurs in the
summer from June onwards. The plant reproduces both
vegetatively and by seed, but little or no seed is generally
produced. When seeds are formed, they are capable of
remaining alive in the soil for many years. The stems
spread the plant by creeping, both above and
underground. Aerial stems can grow back into the soil
in late summer to form new creeping underground stems.
These underground stems are white and brittle and each
piece left after cultivation can produce a new plant.

Control

Where ground can be regularly cultivated, Hedge
Bindweed is readily controlled, but where it cannot, as
among fruit bushes, ornamental shrubs and in hedges,
the weed can be a great nuisance. Cutting the plant down
only results in more stems being produced. It is quite
susceptible to weedkillers containing MCPA or 2,4-D, but
these will have to be applied to the leaves with a
paintbrush when it is growing among ornamentals.
Paraquat/diquat is also effective where it can be used.

JAPANESE KNOTWEED

Description

Japanese Knotweed (*Polygonum cuspidatum*) is an erect and attractive perennial plant growing to over 9 feet (3 metres) tall. The stems look like very thick bamboos and are usually flecked with purplish-red marks. The leaves are big and broad and the cream-coloured flowers are borne in large decorative clusters. Below ground are stout creeping stems, which vary greatly in spreading distance. Their tips turn up in spring and give rise to new aerial shoots. This *Polygonum* is the commonest of the large planted ones and can be distinguished by the leaves, as the bottom edge goes more or less straight across at right angles to the stalk.

Occurrence

Japanese Knotweed was first introduced into this country as an ornamental in about 1825. It became popular and was widely planted and is now naturalised over most of the British Isles, occuring quite frequently in some parts of the west and in the Greater London area, for it tolerates air pollution. It is said to be spreading rapidly in the south west of England.

Reproduction and Spread

In established plants, the new shoots appear above ground in late February or early March and grow up very rapidly indeed. The aerial shoots die back again during November. New shoots appear further outside the clump each year as the underground shoots move gradually outwards and extend the colony. Flowering occurs between July and October. The amount of seed produced is probably not very large, but viability is good after a period of dormancy.

Control

This plant is easy to establish, but difficult if not impossible to eradicate. Containment, by digging up the outside shoots of the clump, or suppression, by cutting down all the shoots frequently, are probably all that can be done. The plant is generally unaffected by herbicides, although glyphosate may have a limited effect; see page 89 at the end of the book.

LESSER CELANDINE

Description

Lesser Celandine (*Ranunculus ficaria*) is a not
unattractive perennial plant, which grows to 10 inches
(25 cm) tall. The heart-shaped leaves are shiny and dark
green. The flowers, which are borne singly, are a bright
glossy yellow in colour. The roots are numerous and are
intermingled with a number of elongated tubers about
1 inch (2.5 cm) long. There are two forms of the plant, one
of which produces bulbils in the angles of the leaves,
while the other produces only seeds. Like other members
of the family, Lesser Celandine is poisonous.

Occurrence

The form producing bulbils is the one found usually as
a garden weed. It appears to prefer shady places. The
other, the commoner form, is less often a garden weed
and appears to prefer sunnier positions. Besides gardens,
Lesser Celandine occurs in various other habitats,
especially grassland and grassy places and also in
woods, hedgerows and by streams, for it prefers damp
situations. It is common throughout the British Isles.

Reproduction and Spread

The flowers open from March to May. The more frequent
form produces seed, but the other rarely (if ever) does
so. The one producing bulbils (which are in fact short
roots with a shoot bud on top) relies on them more or
less completely for propagation, and they can readily give
rise to new plants. Both forms have root tubers, which
can also give rise to new plants if broken off the parent
plant.

Control

As a weed this plant is easily uprooted, but fragments
of tuber are often left behind and these will produce new
plants. Disturbance also helps in dispersing the bulbils,
which also result in new plants. Therefore, it needs
extreme care to dig it out completely. Lesser Celandine
is quite susceptible to 2,4-D and MCPA so it can be
controlled in lawns; if growing among ornamentals, the
weedkillers must be painted on the leaves.
Paraquat/diquat is very effective on this weed, where it
can be used.

SPRING BEAUTY

Description

Spring Beauty (*Montia perfoliata*) is an annual weed
growing to about 1 foot (30 cm) in height. The leaves,
which are long-stalked and rather fleshy, are grouped in
a rosette. Each leaf stalk curves gracefully upwards. Two
other leaves form a cup beneath the white flowers, which
are borne on short stems arising from the rosette.

Occurrence

This is a weed of increasing importance; since its
introduction in 1852, it has become widely distributed
and is still spreading. It occurs mainly in the south and
east of England, where it has become abundant in some
localities. It occurs mainly in cultivated ground on light
sandy soils, particularly in horticultural nurseries,
gardens and waste land.

Reproduction and Spread

Seedlings appear in the autumn and the plant
overwinters as a leafy rosette. Flowering takes place in
late spring and in summer. When ripe, the seed capsule
ejects the seeds violently to a distance of several feet.
Seeding must therefore be prevented completely,
otherwise the garden will be quickly overrun by the weed.

Control

Control of Spring Beauty is best achieved by hoeing, but
hand pulling is possible when only a few plants are
present. The effectiveness of many herbicides is
unknown, but the paraquat/diquat mixture and 2,4-D,
where they can be used, will give useful control when
the plant is young; see page 81 at the end of the book.

CHICKWEED

Description

Chickweed (*Stellaria media*) is a soft, limp, light green, annual or overwintering weed with white flowers. It grows flat on the ground, often forming extensive well-branched mats. Under good growing conditions the weed can become a lush entangling mass a foot or more thick.

Occurrence

Chickweed grows in cultivated land and waste places on all soils, especially rich ones. It is certainly one of the most important weeds of gardens, where it occurs mainly in flower and vegetable beds. It is ubiquitous throughout the British Isles and is cosmopolitan in distribution.

Reproduction and Spread

The seeds, like those of many serious garden weeds, can germinate in all months of the year. The plant grows quickly, has a short life span and can flower and ripen seed the whole year round. It can withstand frost. Up to three generations can occur in a single year and, on average, each plant produces more than 2000 seeds. The seeds can germinate as soon as shed or remain dormant in the soil for many years. Viability can be retained for up to 40 years.

Control

Control by hoeing is possible with young plants in dry weather, as the plant relies initially upon the single central tap root; later on, when roots are formed along the spreading stems, it becomes impossible to hoe out and, in wet weather, the plant will root itself again anyway. Hand pulling is never possible as the stems are too brittle. Fortunately the plant is susceptible to the paraquat/diquat mixture, to dichlobenil, chloroxuron, propachlor, simazine, 2,4-DES and chlorpropham. See Chapter 6 on weed control on page 81 at the end of the book.

CLEAVERS

Description

Cleavers (*Galium aparine*) is an annual weed which usually scrambles over other plants. The stems are flexible but not strong and, like the leaves, have many hooked bristly hairs on them so that they cling to clothing. The flowers are whitish in colour and inconspicuous. The large seeds are circular, and also hooked and clinging.

Occurrence

Cleavers is frequent in farmland and waste places, especially hedges, and can be a problem in gardens. It germinates in autumn and winter at a time when most gardeners are hibernating, so that when spring comes, it is well established and can grow on quickly. It occurs throughout the British Isles.

Reproduction and Spread

As an annual weed, Cleavers relies on seed and it produces plenty. As the seeds are covered in hooked bristles, they are easily dispersed on clothes and on the fur of pets, so seeding should always be prevented.

Control

The hoe is probably the best way of controlling this weed. It is difficult to hand pull for the stems readily break and then cling obstinately to you and other plants. It is susceptible to mecoprop, but Cleavers rarely grows in places where this can be used. It is also susceptible to propachlor and dichlobenil. See page 81 at the end of the book for uses.

MOUSE-EAR

Description

Mouse-ear (*Cerastium holosteoides*) is a trailing slender perennial weed. It is much branched and the small leaves are in opposite pairs. The leaves are densely covered with erect silvery hairs. The flowers are small, white and inconspicuous.

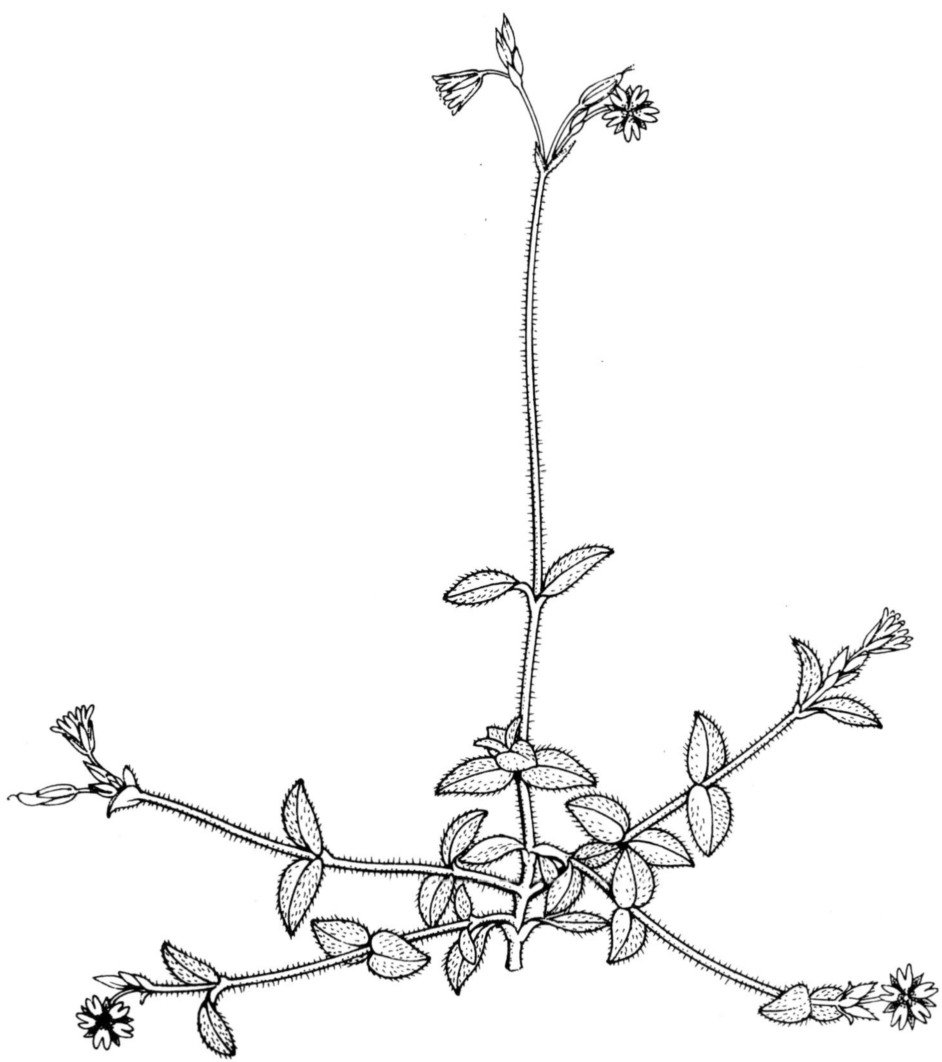

Occurrence

Mouse-ear grows widely in farmland, grassland, roadsides and in gardens, where it occurs mainly in lawns. It is very common and is distributed throughout the British Isles.

Reproduction and Spread

This weed relies on seeds for dispersal and several thousand may be produced by each plant. Flowering occurs from spring to autumn. The seeds germinate mainly in the autumn and only a few in spring. They are long-lived in soil and can survive for up to 40 years.

Control

It is easy to control in lawns, where it can form unsightly patches, because it is generally susceptible, especially when young, to mecoprop and dichlorprop and moderately susceptible to MCPA and 2,4-D, which are used in various combinations in many weedkillers for lawns. In cultivated ground, it is easiest to hoe it out. On paths and drives, it can be controlled by simazine or dichlobenil.

RIBWORT

Description

Ribwort (*Plantago lanceolata*) is a perennial weed of grassy places. It forms a rosette of long narrow leaves, which grow up at an angle. They are very different from the broad flat rosettes of the two other closely related Plantains. The brownish flowers are borne in a cluster at the tip of short unbranched stalks.

Occurrence

Ribwort can occur in a variety of situations, such as grassland, waste places and gardens. Although, unlike the other Plantains, it doesn't grow in close cut lawns; it prefers long grass. It is frequent throughout the British Isles.

Reproduction and Spread

Growth of Ribwort starts in spring and flowering occurs from April to August. Seed production varies with the situation, but an average of 2500 per plant per year has been recorded. The seeds are thought to be able to survive for several decades in the soil.

Control

This weed is not of great significance as a weed in long grass, but regular close mowing is an effective way of reducing it, if its control is required. Alternatively, it is susceptible to 2,4-D, MCPA, mecoprop and dichlorprop, so spraying with mixtures of these (as ingredients in many herbicides for lawns) will kill it, but avoid nearby plants you wish to keep.

PROCUMBENT PEARLWORT

Description

Procumbent Pearlwort (*Sagina procumbens*) is a very
small perennial weed with narrow leaves. It forms small
patches of leaves and stems, which may extend laterally,
especially in trampled areas. The central rosette gives
rise to creeping and rooting flowering stems. The minute
flowers have small white petals, or even none at all. A
similar weed, the Annual Pearlwort (*S. apetala*), also
occurs on paths and lawns.

Occurrence

Procumbent Pearlwort grows commonly on paths, in
grass and on lawns, particularly closely-mown ones,
where it can form quite large patches. It is frequent on
paths for it is resistant to treading. A common plant, it
occurs throughout the British Isles.

Reproduction and Spread

The plant flowers between May and September. The
seeds are small, on average 70 or more are produced
per capsule, and the flowers can be borne in profusion.
The seeds are so small that they can be dispersed by
the wind. Germination occurs almost at once as there
appears to be no dormancy.

Control

Control of Pearlwort in lawns can be achieved with any
of the weedkillers for lawns. Re-establishment is,
however, likely to occur from seed. On paths, dichlobenil
or mixtures containing simazine should be suitable.

ANNUAL MEADOW-GRASS

Description

Annual Meadow-grass (*Poa annua*) is a small annual grass, which grows with its shoots either upright or growing outwards at an angle. It can grow to 12 inches (30 cm) tall. The leaves are generally short, often curved and frequently somewhat crinkled. The flowering head is spreading and triangular in outline. The flowers are green or sometimes tinged with purple. It is variable in shape and size.

Occurrence

Annual Meadow-grass is one of the commonest weeds of gardens. In flower and vegetable beds, it can occur as a continuous green sward and is consequently a serious pest. On the other hand, it can be useful in lawns, because of its continuously green appearance and its ability to grow where other grasses will not, although the flower heads may be considered unsightly by some. It grows in most situations and is virtually ubiquitous in the British Isles.

Reproduction and Spread

The seeds germinate all the year round and the plant can certainly flower and seed in every month also, but the seeds are mainly produced between April and September. It can have, therefore, several generations in the course of a single year. It is very hardy and can withstand severe frost.

Control

Control by hoeing is generally tedious because the fibrous root system is hard to cut through. In addition, plants left on the soil surface will root again in wet weather. Hoeing is easier in cold weather in early spring, for at this time the roots die back considerably. The herbicides, dalapon, dichlobenil, the paraquat/diquat mixture and simazine will all control this weed on paths and drives and in established shrubberies. Propachlor, chlorpropham and paraquat/diquat can be used on cultivated soil under certain conditions; see page 82 at the end of the book.

BARREN BROME

Description

Barren Brome (*Anisantha sterilis*) is an annual grass which grows to 3 feet (90 cm) tall. The plant is usually a light green in colour, but is tinged with purple after flowering. The leaves are flat and softly hairy. The flower head is long-stalked and drooping, with long florets, which are grouped into spikelets (clusters). The seeds, which fall off when ripe, are also long, up to about ½ inch (1 cm), and there is a very long bristle at the end.

Occurrence

Barren Brome is a very common plant and grows on most soils, but especially on light, sandy and well-drained ones. It occurs very frequently by the sides of roads and in waste places, but it can also be a serious nuisance in gardens and on farms. Barren Brome grows throughout the British Isles, but is much less common in the north and west than in England.

Reproduction and Spread

As an annual weed, this plant, despite its name, produces many seeds which germinate in the autumn soon after shedding. The seedlings grow quickly, for the seeds are large, and flowering occurs the following year between May and July. The seeds are rough and long and are often dispersed by catching on clothing or animal fur.

Control

The only answer to this weed in cultivated soil is the hoe, for it is resistant to most herbicides and hand pulling is difficult because the stems break easily, leaving the roots in the soil to regenerate. If it is allowed to get to flowering, it is best to burn the plants, otherwise seed may yet be formed. On paths and drives and in established shrubberies, simazine, dichlobenil, paraquat/diquat or dalapon can be used.

COUCH-GRASS

Description

Couch-grass (*Agropyron repens*) is a very vigorous perennial grass with creeping underground stems, which are white and generally straight. This is probably the only grass with creeping underground stems which occurs frequently in gardens. It occurs as clumps of stems or as odd scattered and rather spiky-looking shoots.

Occurrence

Couch-grass is one of the worst perennial weeds in this country and is certainly widespread and often very common in agricultural land, gardens, horticultural nurseries, waste places, roadsides, hedgerows and elsewhere throughout the British Isles, excepting some highland areas in Scotland.

Reproduction and Spread

New aerial shoots emerge in autumn or early spring. They arise from the tips of the underground creeping stems or from the small buds scattered along their length. These buds, which remain dormant when the tip is growing strongly, can produce new aerial shoots very rapidly when the underground stems are cut up by cultivation. The leafy shoots give rise to new

underground stems from buds just below the ground, during late spring and summer. Some of the aerial shoots can produce flowers, but seeding is unlikely in gardens except in neglected areas. The seeds will germinate at once.

Control

Control can only be by constant hoeing or by digging up the wiry subterranean stems. In open soil this is quite easy, for the stems never run very deep, but when it is among the roots of ornamentals it can be very difficult to eradicate. There are several herbicides which will give control; see the table of susceptibilities on page 87. It is useful to hoe or treat the shoots with paraquat/diquat in late summer to prevent or reduce the production of new underground stems.

5. HOW TO CONTROL WEEDS

When you have determined which weeds are present and which are most serious and needful of control, then there is the question of how to control them.

There are more ways of killing a weed than just laboriously digging it out with a spade. The choice of method should be determined by various factors such as, the characteristics of the weed, the number of plants involved, the size of the area they occupy, the presence of plants you wish to keep, the implements available, etc. For the average gardener there are seven main alternatives for weed control: hand pulling, hoeing, digging out, mulching, mowing, flaming and using herbicides.

Hand Pulling
This is probably one of the most widely-used methods and is certainly a feature of the evening stroll to the bottom of the garden, when one is so often faced by a tall weed with a single stem growing, hitherto unnoticed, amongst some prized ornamental or in an otherwise perfect row of vegetables. The human response is inevitable. However, the efficacy of this response depends very much on the weed itself. Those plants which lend themselves to hand pulling are generally annuals, invariably tough stemmed (or they will break off), shallow rooted and tall. In the table on page 85 it is indicated whether it is worth while trying to pull up the weeds listed in this book. About one-third of those on the list are suitable, provided that there are not too many together; the most suitable are Canadian Fleabane, Broad-leaved Willow-herb, Fat Hen, Feverfew, Fool's Parsley, Nipplewort and Poppy.

Having pulled the weed successfully, do not then drop it out of sight on to the soil surface, for it could re-root if the soil is damp. Make use of the weed by putting it on the compost heap or, if it is about to seed, burn it and scatter the ash on the garden to fertilise the plants you want to grow.

Hoeing
The hoe is a much-maligned instrument, possibly because it seems to conjure up the idea of back-breaking, monotonous labour, but it is nonetheless a most effective method of controlling weeds.

There are basically two types of hoe – those you push, called the 'Dutch hoe', and those you pull or chop; there are some that are capable of both actions. The object of all types is to cut the weed just below the soil surface so that the stem is separated from the root. To accomplish this, you need a sharp blade as well as a strong wrist. The greatest effect is achieved by dividing the plant *exactly* into stem and root. If cut too deep, the aerial stem will have some root attached, so it can possibly re-root; and if cut too high, the root will have a length of stem attached from which it can produce new shoots. There is also the possibility that, if the hoeing results in too much soil disturbance, dormant weed seeds lying in the soil will be disturbed and thereby encouraged to germinate. Another flush of weeds will then have to be dealt with.

Hoeing is always most effective if performed on a drying day, the weeds then quickly dry out and die. Similarly, it is also most effective if performed when the seedlings are small, which is also when least effort is required to cut the stem. If the weather is wet, then it is worth raking the weeds up and composting them.

Many gardeners are happy with a hoe alone and much can indeed be done to keep the weeds in order with this implement, although in some soils, especially heavy or stony ones, this may be more difficult. It may be difficult too to use hoes in rockeries or on paths of uncemented brick or stone. For some weeds in open soil, especially creeping perennials, hoeing does little to effect control, and the use of fork or herbicide is often the only really satisfactory solution.

Digging Out
Digging with a spade is a useful way of sweeping the weeds under the carpet at the same time as preparing the soil for the next lot of flowers or vegetables. Turning the soil over is a very useful way of getting rid of weeds, but should be restricted to annual weeds before they have produced any seeds; with perennials, one is simply putting off the day when the problem has to be faced – indeed delay usually makes the problem worse. Cutting up a creeping perennial weed into sizeable lengths only makes more plants; there will be two or more shoots where only one grew before, admittedly, the shoots will be small, but they will grow. On the other hand, the garden fork can be used to remove creeping perennials from light soils, but only those that have shallowly-creeping parts and are not too brittle, otherwise the problem proliferates. Where forking is possible; weeds such as Couch-grass, Creeping Bellflower, Winter Heliotrope, Blue Sow-thistle, Rosebay Willow-herb, Ground Elder, Stinging Nettle, Creeping Buttercup and Enchanter's Nightshade can be removed, although the last is rather brittle. Tap-rooted weeds, such as Docks and Dandelion, can be similarly treated.

Having dug out the underground parts with great care, do not put them still alive on the compost heap, for they will quickly establish themselves there. They must be dried out first or burnt.

Mulching
The practice of mulching, that is the spreading of leaves, compost or other materials around newly transplanted plants to prevent them drying out, has occasionally been adapted to suppress weeds. Materials as diverse as newspaper and pine needles have been tried over the years. A modern development has been black polythene, which is occasionally used in strawberry beds, where it conserves moisture, raises the soil temperature and encourages weed germination. The weeds that germinate soon exhaust the food stored in the seed and then die, for there is no light available. Because black polythene is not very attractive, it is probably best restricted in the garden to crops, such as strawberries, which are difficult to weed. It can also be used at the back of a shrubbery where it cannot be seen. Other materials, unless put on particularly thickly, have only a temporary effect.

Mowing

Mowing is not a very useful method of weed control. In a new-sown lawn, many annual weeds will grow up with the grass and an early high cut will prevent them from producing seed, but will do little to get rid of them. Couch-grass growing up from underground stems in this situation is, however, usefully controlled by repeated mowing. So there is no need to worry too much about its presence when sowing a lawn.

Perennial weeds in established lawns, such as Yarrow, Daisy, Dandelion and Plantains, can be prevented from seeding by mowing, but they will persist for year after year no matter how closely the grass is cut. They are best treated with a herbicide. Other weeds of lawns, such as Black Medick, Lesser Trefoil, Mouse-ear and Annual Meadow-grass take benefit from mowing, for their seeds are distributed ever more widely about the grass.

Flame Guns

A method of killing weeds on paths, drives, along fences, in crazy paving, and on bare soil is by flame gun. These run on paraffin under pressure and burn with a hot blue flame. They are rather slow and naturally consume quite a lot of fuel, but are effective, especially on young seedlings. Larger plants, of course, will take a relatively long time to incinerate, particularly if they are succulent. Flame guns are very useful in situations where weed seedlings are emerging among dead plant material, as in borders of herbaceous perennials, for the bed is cleaned of weeds and rubbish in one go and no disturbance of the ornamentals is necessary. Flaming will not kill weed seeds in the soil.

Another place where flame guns can be extremely useful is along wire fences; in early spring, the flame rapidly burns away all the old dead stems entangled in the wire, a job which would take a lot of time if tackled by hand.

Herbicides

The use of herbicides or weedkilling chemicals is perhaps a rather novel method of gardening, but it is one which offers an easy solution to many of the intractable weed problems that have beset gardeners over the years.

Many people do not believe in the use of these chemicals, possibly because they do not know much about them or are scared of their almost magical properties and believe that they must be poisonous to humans too. However, a significant proportion of gardeners do use them to some extent and they have derived considerable benefit. Herbicides do present problems, mainly through their remarkable effectiveness, so it is essential to use them sensibly and with care, for the greatest risk is that you will kill your flowers and vegetables as well as the weeds. Damage to your neighbour's plants is of course something that is to be avoided even more strenuously. Make sure that you find out all you can about a weedkiller before applying it and make sure that you follow the recommendations exactly, then you should have no problems.

The main question about herbicides is whether it is really worth using them. The answer depends partly upon the size of the area involved and partly upon whether the herbicide really does provide an easier solution than alternative methods. On lawns, where hoe and fork are largely irrelevant, the prospect of hand weeding is not good unless only a few specimens of an easily weeded plant, such as Lesser Trefoil, are present. So, for a majority of lawns, herbicides provide an easier solution irrespective of size.

For paths and drives, if small and not very weedy, a hoe is the simplest cure unless the material is hard and well-bedded gravel, but if large and weedy then a herbicide can provide a very much easier solution.

For cultivated soil, the size of the area is probably one of the more important considerations. The soil is disturbed and vegetables, at least, are grown in straight spaced rows, for ease of hoeing; so why not hoe? However, if the area is large and complicated by fruit bushes (where frequent access may be difficult or damaging) or if growing is on a large scale for self-sufficiency and some extra for sale, it could be that herbicides will reduce the time and the labour involved in controlling the weeds, especially if a single application in early spring is all that is necessary. The outstanding problem of flower and vegetable beds is the creeping perennial weed which it is often just not feasible to dig out. It may then be worth using the rather slow method of painting individual leaves with a small paintbrush loaded with an appropriate herbicide, rather than excavating.

Other Possibilities

Another method is to reduce the chances of weeds getting established by planting competitive ornamentals to exclude them. This can be quite effective as far as annual weeds are concerned. Fill up the soil with roots of a vigorous ornamental and shade the ground with its leaves, and this will do much to discourage annual weeds. It is rather a long-term method of weed control and not necessarily always effective.

There are also those who believe that by planting particular species, which are thought to have some sort of toxic exudate produced in the roots, many of the garden weeds can be effectively reduced or contained. This method has not as yet been satisfactorily proved.

6. WEED CONTROL IN VARIOUS SITUATIONS IN A GARDEN

Newly-sown Lawns

A newly-sown lawn is very prone to weed problems in the early stages, as the disturbance caused by levelling will have stimulated many of the weed seeds in the soil to germinate. As soon as the grass seeds start to germinate, so also will the weeds, if not sooner. Grass seedlings in the earliest stages are not very competitive and the weeds with big broad leaves appear to leave them behind. Unfortunately, most lawn weedkillers, which includes MCPA, 2,4-D, mecoprop, dichlorprop and others, cannot be applied to fine-leaved grass mixtures before about six months after sowing because the grasses will not tolerate them. Fortunately, ioxynil has proved safe on young grasses and can be applied when they have two fully-expanded leaves. Make sure that you get the formulation designed for newly-sown lawns and not the one with mecoprop added.

Established Lawns

There are a large number of different plants, other than grasses, which can grow in lawns. The majority are relatively inconspicuous and harmless and merely detract to a limited extent from the neatness of the lawn. There is a small hard core of lawn weeds which are either aggressive or are particularly noticeable and so are more definitely weed plants. There are about a dozen of these mentioned in this book.

Weeding lawns by hoe or fork is not really possible without doing a lot of damage and weeding lawns by hand is best left to the enthusiast, so it is fortunate that a number of herbicides have been discovered which will kill only the 'broad-leaved' weeds without harming the grasses. This is why there are so many weedkillers for lawns. Most are based on MCPA or 2,4-D, which kill many of the lawn weeds, such as Dandelion, Daisy, Mouse-ear, Plantains, Thistles, Creeping Buttercup etc. There are some weeds which are less susceptible, such as Yarrow and Procumbent Pearlwort, and others which are completely resistant, such as Slender Speedwell, White Clover, Black Medick and Lesser Trefoil. To deal with these resistant weeds, a number of other herbicides such as mecoprop, dichlorprop, fenoprop, dicamba, etc., are added to 2,4-D or MCPA to make a complete weedkiller. Ioxynil can be used on newly-sown lawns as a general weedkiller or with mecoprop on established ones to kill Slender Speedwell as well as the others. Another complete treatment is mecoprop with bromofenoxim added and there may be others. Tar oil is also available for controlling Slender Speedwell and mosses. Most lawn weedkillers are applied to the whole lawn as a liquid by spraying or watering (for methods, see under How to Apply Chemicals on page 88). For small lawns or a few weeds in a large one, aerosols with a foam to show which weeds have been treated, and hand-held applicators for touching individual weeds, are available to save overall treatment.

One of the most difficult problems in lawns is other grasses. Some look weedy because they are a different colour, such as Yorkshire Fog which is hairy and light grey-green, while others grow out rapidly from the edges of lawns, such as Creeping Bent, and constitute a problem in areas surrounding the lawn. These cannot be killed selectively out of the acceptable grasses by herbicides. The only remedies are to dig out the offending grass or to kill the grasses locally with a herbicide. Both methods will entail subsequent re-seeding of the area unless it is on a very small scale.

Acceptable grasses will produce flower stems, which can also be a problem. These are often tough and difficult to cut down with the conventional grass mower. A machine must be used which has a horizontal cutting action, or the stems cut by hand with sickle or shears.

Mosses

Mosses are frequent in lawns and elsewhere, and there are a number of weedkillers to control them. Lawn sand, which is ammonium sulphate and ferrous sulphate, kills moss and is also effective against a number of lawn weeds. Calomel or mercurous chloride, chloroxuron, dichlorophen and tar oil are also available for moss control. For more permanent results, it is perhaps better to alter the conditions, for moss is frequently symptomatic of excessive moisture. Improve the drainage or reduce the shade that conserves moisture and there should be a general improvement.

Rough Grass

In areas of rough unkept grass, many of the weeds of mown lawns will also be found, as well as Docks, Thistles, Stinging Nettles, Brambles and woody weeds. Treatments for these areas will depend on whether it is the grass, bushes, trees, or other plants that are to be preserved. For total kill of vegetation, sodium chlorate can be used. This is more suitable for a large garden, but even there it must not be used where it can move laterally through the soil as on a slope, nor used above the spreading roots of trees and shrubs if they are to be preserved. It will persist in the soil for up to a year afterwards. Ammonium sulphamate, which persists for about three months, can be used as an alternative. If the area is large, it might be advisable to get a contractor to plough it.

For selectively killing out the broad-leaved weeds from the grass, many of the lawn herbicides are suitable. If there are bushes to be killed, then there is 2,4,5-T, with or without 2,4-D; 2,4,5-T is the most effective herbicide for killing shrubs and trees. It is probably easiest to cut down the unwanted bush first and then apply 2,4,5-T to the stump.

Flower Beds

In established herbaceous borders, propachlor granules can be applied to clean soil or chloroxuron can be watered over clean-weeded soil to kill annual weeds as they germinate. 2,4-DES in a mixture with simazine and chlorpropham mixed with propham and diuron are also available for controlling annual weeds. If the weeds are already established, then hoe or hand-weed (see the table on page 85 for feasibility), or use the paraquat/diquat mixture dribbled on to the weeds through the dribble bar sold for this purpose, but avoid dribbling it on to ornamentals. When the soil is free of weeds, then apply one of these residual herbicides to keep it clean. Around bedding plants, herbicides such as chloroxuron can be applied to clean soil within 48 hours of planting out. Always make sure that the product you buy is suitable for flower beds and for the type of plants growing there.

Perennial weeds in bedding plants or herbaceous borders are rather less easy to control, especially if you cannot dig them out. Many of them are susceptible to the weedkillers for lawns, but so too are the ornamentals. A drop or two of weedkiller can be painted on to the central vein of two or three leaves on each shoot with a paintbrush. Great care must be taken not to drop any on to desirable plants.

Shrubberies
Simazine can be applied to clean weed-free soil under a wide range of ornamental trees and shrubs provided they are well established. Applied in early spring, the herbicide will control annual weeds for the whole season. Dichlobenil can also be used to control annual weeds and a number of perennial weeds too. Newly-planted or small shrubs should not be treated with these herbicides, nor should herbaceous plants, and there is an added risk of injury on light soils which contain little organic matter. If in any doubt about the safety of shrubs or trees, consult the manufacturer. Other residual herbicides, which will control germinating annual weeds for varying periods of time are propachlor, chloroxuron, chlorpropham with propham and diuron, and 2,4-DES with simazine. Dalapon can be used if Couch-grass is present.

Some of the residual herbicides will not persist for the whole season and may need to be re-applied, so consult the manufacturers' recommendations. If a few weeds appear later on, treatment of individuals with paraquat/diquat can be useful, for the soil is not then disturbed, which would encourage further weed germination. Paraquat/diquat should be kept off foliage and young bark of ornamentals.

When planting a shrubbery, it is worth grouping plants that are tolerant of particular herbicides to ensure easy and trouble-free weed control.

Fruit Bushes
Weeds growing among fruit bushes need to be treated very early in the year. Start by killing any existing weeds by using a hoe or by dribbling paraquat/diquat mixture over them. Then, when the soil is clean, apply one of the residual herbicides to keep it clean. These are simazine, 2,4-DES with simazine, dichlobenil, chloroxuron, propachlor and chlorpropham with propham and diuron. These can be used on most tree and bush fruits, but check whether the product to be used is suitable for the situation and whether it is subject to any special restrictions. This applies particularly to the time of application. If perennial weeds are present, dichlobenil should be used, but the fruit trees and bushes must be well established (i.e. planted at least two years previously).

It is usually best not to disturb soil under bushes more than is necessary, because every disturbance will encourage more buried weed seeds to germinate. Similarly, having applied simazine or other residual

herbicide, do not hoe, otherwise you will make holes in the continuous layer of chemical on the surface and allow weeds to establish. To avoid disturbance it is also sensible to choose the longer-persisting herbicides, such as simazine or dichlobenil.

Vegetable Beds
Vegetable beds are usually laid out in straight spaced rows, which enables that most useful weapon, the hoe, to be used. In most gardens, the small size of the vegetable patch means that nothing more is needed, but in large gardens and when creeping perennial weeds are present, then a herbicide often has to be used.

Herbicides used to control annual weeds in vegetables are chlorpropham (in mixture), chloroxuron and propachlor; 2,4-DES with simazine can be used in a few crops. Make sure before using any of these that all the vegetables are tolerant of them. These herbicides must be applied to weed-free soil. They will keep it clean for varying periods of time depending on the chemical and the conditions. It is best not to hoe, but if really necessary, hoe little and lightly. Alternatively, the paraquat/diquat mixture can be used, but be careful to avoid the vegetables. This mixture can also be used in the preparation of weed-free soil. Black polythene can also be used, with edges buried, between rows of vegetables to prevent weed establishment, but slug pellets should be put underneath.

If creeping perennial weeds are present, then it is best to paint two or three leaves on each shoot with a paintbrush loaded with one of the lawn herbicide mixtures, but be careful to avoid treating the crop too. The hoe persistently used, or the paraquat/diquat mixture frequently applied, will also discourage them.

Annual Weeds in Cultivated Land
There are three main methods of controlling annual weeds in cultivated land: hoeing, digging or using a residual herbicide, which sits on the surface for one to several months and kills each weed as it germinates. Paraquat/diquat can be used to kill the young seedlings but it does not persist. The suitability of hoeing and digging is given in the table on page 85. The use of residual herbicides depends on the situation – see Flower Beds, Shrubberies, Fruit Bushes and Vegetable Beds above. The table gives guidance on the most suitable herbicides to use. The entries indicate particularly effective treatments, so the absence of a herbicide does not necessarily indicate lack of activity.

Perennial Weeds of Cultivated Land
These can be some of the worst problems of gardens. Enquiries from gardeners have shown that the worst weeds are Ground Elder, Couch-grass, Slender Speedwell, Horsetail, Field Bindweed and Pink-flowered Oxalis, in that order.

Hoeing is one way of suppressing them, but it must be done frequently and rarely eliminates them. Digging out of some perennials is possible (see the table on page 85), if one can get at them. Many are susceptible to

herbicides for lawns, so it is suggested that they be
treated by painting two or three leaves per shoot with
a paintbrush dipped in herbicide. Great care must be
taken to avoid touching desirable plants.

Rose Beds
Roses are often grown alone in beds, which makes
control of annual weeds easy. Residual herbicides such
as chloroxuron, chlorpropham (in mixture), dichlobenil,
2,4-DES with simazine, propachlor or simazine are
suitable if applied to weed-free soil. Dichlobenil will
control perennials too (see table on page 85) but should
only be used among established roses. It should be
applied early in the year.

*Weeds on Paths, Drives, Courtyards and Hard Tennis
Courts*
The absence of other plants in these areas makes control
much easier. For total weed control, sodium chlorate can
be used but it shouldn't be used where it can move, as
on slopes, nor over tree roots. Many path weedkillers are
based on simazine and have others added, such as
aminotriazole, MCPA or paraquat/diquat to give both a
quick and prolonged effect. Dichlobenil will also give
prolonged effect. Paraquat/diquat alone will give general
weed kill, but will need re-applying later. Tar oil
preparations are available, which will also kill mosses.
For grasses, dalapon can be used.

7. TABLE OF WEEDS AND THEIR SUSCEPTIBILITIES

The Table
The purpose of the table is to indicate which treatments are most effective in controlling individual weeds in particular situations. The table is not complete, partly to keep it simple and partly through lack of knowledge. Some general treatments have been omitted, for example, sodium chlorate. Others are only mentioned occasionally, usually when they are particularly useful. Reference to the uses of the chemicals can be found in Chapter 6: Weed Control in Various Situations in a Garden on page 81, and in the list of chemicals on page 89 at the end of the book. Any information that will make the table more useful in subsequent editions will be welcomed.

The table is full of possibilities for control by herbicides, but this does not imply that herbicides are invariably the best treatment. If you do decide that a herbicide is necessary, use only suitable products, and in the place and manner described in the manufacturers' instructions. Do not use herbicides that are not for gardens.

Most of the table is self-explanatory. The column marked 'Feasible to Pull, Hoe or Dig' is designed to give an indication of the possibility of achieving useful results by these methods, but obviously the possibilities will vary. Entries against an individual weed are made only for situations in which it frequently occurs. The order of listing the treatments is no indication of order of efficacy, nor is the absence of a treatment any indication that it will not be effective on that weed.

The list of weeds in the table follows the same order as the illustrations, which are in a sequence based on leaf outline.

Since this book was written, glyphosate has become available to gardeners. It will kill most plants and is particularly effective in controlling weeds with creeping underground stems. Conversely, chloroxuron is no longer available, so references to this chemical should be ignored.

TABLE OF WEEDS AND THEIR SUSCEPTIBILITIES
(Where several chemicals are listed, these are alternatives)

Picture on page	Weed Name	Annual or Perennial	Creeping Weeds Roots or Shoots	Shallow or Deep	Seeds Produced	Feasible to Pull	Feasible to Hoe	Feasible to Dig	Lawns (1)	Paths & Drives (2)	Vegetable & Flower Beds	Shrubs & Bushes (3)
19	Rayless Mayweed	A			Yes	P	H			Sm, Db	Cx, Pc	Sm, Db
20	Horsetail	P	S	D	None		H			Db	Hoe, Dig	Db
21	Yarrow	P	S	S	Yes				Dc, Mc, Fp, Dm			
22	Fumitory	A			Yes	P	H			Sm	Hoe, Ds, Ch	Sm
23	Black Medick	A			Yes				Mc, Dc, Dm, Fp			
24	Lesser Trefoil	A			Yes	P			Mc, Dc, Dm, Fp			
25	White Clover	P	S	Surface	Few				Fp, Mc, Dc, Dm			
26	Pink-flowered Oxalis	P	Bulbils	S	None					(7)	(7)	(7)
27	Yellow-flowered Oxalis	A or P	S	Surface	Yes		H			Sm, P/D, Db	Hoe	P/D, Sm
28	Hairy Bittercress	A			Yes	P	H			Db, Sm, P/D	Pc, Hoe	Db, Sm, P/D
29	Creeping Buttercup	P	S	Surface	Yes		H	D	Any L W	P/D, Sm	P/D, Hoe	P/D
30	Ground Elder	P	S	S	Few			D			Dig, (4)	Db
31	Creeping Yellow-cress	P	R	S	Few		H			Db	Hoe, P/D, (5)	Db
32	Groundsel	A			Yes		H			Db, P/D, Sm	Cx, Pc, Ds	Db, P/D, Sm
33	Nipplewort	A			Yes	P	H				Hoe, P/D	P/D
34	Fool's Parsley	A			Yes	P	H			Db	P/D, Hoe	P/D, Db
35	Shepherd's-purse	A			Yes		H			Sm, Db	Cx, Pc, Ds, Ch	Sm, Db
36	Perennial Sow-thistle	.P	R	D	Yes					Db	(5)	Db
37	Poppy	A			Yes	P	H				Cx, P/D, Ch	Sm, Db
38	Dandelion	P	Tap root		Yes			D	2,4-D	Db	(5)	Db
39	Feverfew	P			Yes	P	H			Pull, Hoe	Hoe, HRU	Pull
40	Creeping Thistle	P	R	D	Few				Any L W	Db	(5)	Db

Table of Weeds and Their Susceptibilities (continued)

Picture on page	Weed Name	Annual or Perennial	Creeping Weeds Roots or Shoots	Shallow or Deep	Seeds Produced	Feasible to Pull	Hoe	Dig	Garden Situations and Control Methods* Lawns (1)	Paths & Drives (2)	Vegetable & Flower Beds	Shrubs & Bushes (3)
41	Sow-thistles	A			Yes		H			Sm, Db, P/D	Cx, Pc, P/D	Sm, Db, P/D
42	Fat Hen	A			Yes	P	H				Cx, P/D, Ds	Sm, Db
43	White Dead-nettle	P	S	S	Yes		H	D			Dig, Hoe	Hoe
44	Annual Nettle	A			Yes	P	H			Sm, Db	Cx, Pc, Ds, Ch	Sm, Db
45	Stinging Nettle	P	S	Surface	Yes			D	Any L W	Db	Dig, (4)	Db
46	Coltsfoot	P	S	D	Yes					Db	(4)	Db
47	Black Nightshade	A			Yes	P	H				Cx, P/D	Sm
48	Gallant Soldier	A			Yes	P	H				Cx, Pc, Ds	Sm
49	Rosebay Willow-herb	P	R	S	Yes			D		Db	Dig	Db
50	Broad-leaved Willow-herb	P	S	S	Yes	P	H			Db	P/D	Db
51	Blue Sow-thistle	P	S	S	None			D		Hoe	Dig, HRU	Hoe, HRU
52	Annual Mercury	A			Yes	P	H				Hoe, P/D	Db, Sm
53	Slender Speedwell	P	S	Surface	None		H	D	Ix + Mc, Ls, To	To	Hoe, Dig	
54	Speedwells	A			Yes		H				Ch, Pc, P/D	P/D
55	Daisy	P			Yes				Any L W			
56	Docks	P	Tap root		Yes			D	Any L W		Dig	Db
57	Red Dead-nettle	A			Yes		H				Pc, Hoe	Sm
58	Henbit Dead-nettle	A			Yes		H			Sm, Db	Hoe	Sm, Db
59	Hoary Cress	P	R	D	Yes					Db	(5)	Db
60	Winter Heliotrope	P	S	S	Rare			D			HRU, Dig	
61	Creeping Bellflower	P	S	S	Yes			D			HRU, Dig	
62	Canadian Fleabane	A			Yes	P	H			Sm	Hoe, Pull	Sm
63	Plantains	P			Yes		H		Any L W	Db	Hoe	Db

*See key and notes at the foot of this table

No.	Weed										
64	Enchanter's Nightshade	P	S	S	Yes		D			Dig, HRU	
65	Sun Spurge	A			Yes		H		Db, Sm	Hoe	Db, Sm
66	Field Bindweed	P	R	D	Few					(5)	Db, (5), (7)
67	Hedge Bindweed	P	S	D	Few		D				(5), P/D
68	Japanese Knotweed	P	S	D	Few		D				R
69	Lesser Celandine	P	Bulbils		Some do		D	Any L W		P/D	P/D
70	Spring Beauty	A			Yes	P	H			P/D	P/D
71	Chickweed	A			Yes		H			Cx, Ch, Ds, Pc, P/D	Sm, Db, P/D
72	Cleavers	A			Yes		H			Pc	Db
73	Mouse-ear	P			Yes			Any L W	Sm, Db		
74	Ribwort	P			Yes			Any L W			
75	Procumbent Pearlwort	P			Yes			Fp, Mc, Dm, Dc, Ls	Db, Sm		
76	Annual Meadow-grass	A			Yes		H		P/D, Sm, Db, Dp	Ch, Pc, P/D	Sm, Db, Dp, P/D
77	Barren Brome	A			Yes		H		P/D, Sm, Db, Dp	Hoe	Sm, P/D, Dp, Db
78	Couch-grass	P	S	S	Sometimes		D		P/D, Dp, Gy, Db	P/D, (6), Dig	Db, Dp, Gy, P/D

*See key and notes at the foot of this table.

Key to herbicide abbreviations

Any L W = Any lawn weedkiller	Dm = dicamba	Ix = ioxynil	R = resistant to garden herbicides
Ch = chlorpropham	Dp = dalapon	Ls = lawn sand	Sm = simazine
Cx = chloroxuron	Ds = 2,4-DES	Mc = mecoprop	To = tar oil
Db = dichlobenil	Fp = fenoprop	MCPA = MCPA	2,4-D = 2,4-D
Dc = dichlorprop	Gy = glyphosate	Pc = propachlor	
	HRU = herbicide response unknown	P/D = paraquat/diquat mixture	

Notes   (1) Where several specific chemicals are listed together, these are alternative treatments, any one of which is suitable. However, most lawn weedkillers are mixtures, so you should buy any product containing the required chemical.

(2) In addition to the chemicals listed in the table, which are suitable for individual weeds, there are also tar oil products and sodium chlorate (where suitable) for general weed control. Simazine is listed alone, but can equally have useful additives.

(3) Simazine is frequently listed for shrubs and bushes, make sure the product is suitable for these situations.

(4) 2,4,5-T painted on the leaves with a small paintbrush.

(5) Any lawn weedkiller containing 2,4-D or MCPA painted on the leaves with a small paintbrush.

(6) Glyphosate (if and when available) painted on the leaves with a small paintbrush.

(7) Oxadiazon (if and when available).

8. HOW TO USE HERBICIDES

Herbicides should always be used with great care, especially in gardens, because flowers and vegetables are very often as susceptible to them as the weeds one wishes to kill – if not more susceptible. There are other hazards too, so one must observe certain general rules.

Safety to Humans and Pets
Most weedkilling chemicals are completely safe to handle but, as a few are not, it is essential to treat all with equal caution. Always read the label and any leaflets first to see if any special precautions are necessary for the safety of humans, pets, fish in ponds, particular plants or groups of plants. Perhaps the greatest danger is the risk of young children getting hold of concentrated chemicals, so these should always be kept out of their reach, preferably in a locked cupboard or shed. *Never* pour the concentrated chemical into another bottle or jug from which it might be drunk by mistake, especially lemonade or beer bottles, for serious accidents have already occurred in this way. *Never* take the watering can containing the chemical into the kitchen to fill it up at the sink; leave the can outside and take the water to it.

Gardeners can, however, be assured that products for use in gardens have been cleared by the Government Pesticides Safety Precautions Scheme for safety to humans, domestic animals and wild life when used strictly in accordance with the instructions on the label.

Safety to Ornamentals and Vegetables
The next most important factor is the safety of plants that you are trying to grow in your garden. Keep chemicals off your clothes, boots and shoes, and always wash your hands thoroughly after using herbicides. When using most herbicides, particularly lawn herbicides, it is advisable to apply them as the last job of the day in order to avoid transferring them by touch on to ornamentals and vegetables, as they have very obvious effects even at exceedingly low doses. Another rule must be that the cuttings from the first mowing of a treated lawn should never be used as a mulch.

How to Apply Chemicals
You should always keep a separate appliance for putting on chemicals. In the average small garden, a watering can is probably sufficient. Paint 'weedkillers only' on the side in case someone else does the watering. Always wash out your watering can very thoroughly before using a different chemical. In a larger garden a pressurised sprayer can be used. If a sprayer is used, no drift or fine droplets should be allowed to fall on plants one wishes to keep or on to a neighbour's plants. The enthusiasm of a watering can rose can be reduced by putting sticking plaster over all but a single row of holes, otherwise one is liable to put the treatment for the whole lawn on to one small corner. In beds with widely-spaced plants, one can treat the weeds growing in between with the paraquat/diquat mixture from a watering can, especially if it is fitted with a special dribble bar sold for the purpose. Great care should be taken not to water ornamentals or vegetables.

Read the label on the container, then follow the instructions carefully. Never add extra amounts for quicker or better results, for some chemicals are actually less effective at higher concentrations. Always mix up the right amount, for disposing of the excess can lead to difficulties. Products are labelled with directions on how to apply them and the area that can be treated. Before diluting the herbicide, it is advisable to make a few tests with a canful of water to get an idea of the speed at which to walk in order to cover the area evenly. It is a good plan to start applying the chemical at the farthest point of a lawn or drive and work from side to side moving backwards towards an exit in order to avoid walking on the treated area and contaminating one's gumboots, which can be as difficult to wash clean as a watering can.

Choosing the most suitable chemical can be difficult. The herbicides in this book are listed by their chemical common name, but the manufacturer will sell his product under a trade name, which is easier to pronounce. Trade names change and new ones are constantly being invented; it would be impossible to attempt to list these in a book of this kind, as it would soon be out of date. Most of the chemicals mentioned here are available in small packs for gardeners, and glyphosate and oxadiazon may soon be. Products have the chemical common names of the ingredients listed on the container label, or in an advisory leaflet supplied with it. This will enable you to pick out one that is suitable for your particular problem weed. However, if you prefer to look for the name of the weed that you have as a problem, see if it listed as being susceptible and, if it is, make absolutely sure that the product is suitable for the area (lawn, path, etc.) where you wish to use it. Some retail outlets may not have all the products available, in which case find a suitable alternative or try another retailer.

Most chemicals available to gardeners have been mentioned, although not in equal prominence. Any omissions or rare mentions are due to efforts to keep the advice straightforward and concise and not to any inadequacy on the part of the chemical or product. Every effort has been made to keep the information simple but useful. No responsibility can be taken for the results derived from any treatment; if the directions on the container are followed, there should be no reason why satisfactory weed control, although not necessarily complete eradication, should not result. To avoid any unnecessary disappointment, it must be emphasised that some chemicals act quickly, while others may take several weeks to reach their maximum effect. So if the weeds do not vanish overnight, do not immediately assume that the treatment is not going to work, for it may be one of the slower-acting ones.

Herbicides used in the garden can come under an official approval scheme. In this scheme, which is voluntary, chemicals are considered for specific uses, and approval is given by the Agricultural Chemicals Approval Organisation if they are satisfied that the product fulfils the claims made on the label. It is in your best interests to choose products bearing the approval mark, which is a large capital 'A' with a crown above it.

Do not be frightened of herbicides, but conversely do not use them indiscriminately. They can be useful, time-saving aids to weed control if applied sensibly.

9. WEEDKILLING CHEMICALS FOR GARDENS

Aminotriazole is occasionally used in weedkillers developed for paths and drives. It is used as an additive to simazine to kill weeds that are already established. It does not persist very long in the soil.

Ammonium sulphamate can be used for general plant kill as an alternative to sodium chlorate. It persists in the soil for up to three months, so the ground cannot be used immediately. It should not be used over the roots of trees, for it kills bushes and trees. It is corrosive to many metals.

Bromofenoxim is available in a mixture with mecoprop to kill weeds in lawns.

Chloroxuron is used for killing weed seedlings as they germinate among flowers and most vegetables. It must be applied to weed-free soil. Its action persists for up to two and a half months. It also kills moss.

Chlorpropham (CIPC) is available in mixture with propham (IPC) and diuron for the control of germinating annual weeds in various crops.

2,4-D is very widely used in lawn weedkillers, for it kills most of the weeds. It is sold with various additives, such as dichlorprop, dicamba, fenoprop and mecoprop to control all the weeds. Some preparations also include fertilisers. 2,4-D is also sold mixed with 2,4,5-T to control a range of large, deep-rooted weeds, such as Stinging Nettles, Thistles and woody plants also.

2,4-DES is available as a mixture with simazine for control of annual weeds in flowers, fruit and shrubs. It persists for up to twelve weeks.

Dalapon is a grass killer and is used especially for the control of Couch-grass. It is applied to the leaves, but is also taken up by the roots. It persists in the soil for several weeks.

Dichlobenil, which is available as granules, kills established annual weeds and prevents both annuals and perennials from becoming established among fruit and ornamental bushes and trees, which must be established (i.e. planted for two years or more). It is only applied in winter or early spring. It must not be used in herbaceous borders, bulbs, seed beds, nor near glasshouses. It persists throughout all the season.

Dichlorprop is available as a mixture with 2,4-D or MCPA to control weeds in lawns.

Dicamba is available as a mixture with 2,4-D to control weeds in lawns.

Fenoprop is available as a mixture with 2,4-D to control weeds in lawns.

Glyphosate is a weedkiller which is not yet available to gardeners, but is mentioned because it is likely to become available. It is particularly effective on Couch-grass. It will also kill a very wide range of other weeds. It is slow to act.

Ioxynil alone is useful for general weed control in newly-sown lawns after the grasses have produced two fully-expanded leaves. It is also available in mixture with mecoprop for control of Slender Speedwell and other weeds in established lawns.

Lawn sand is a mixture of ammonium sulphate and ferrous sulphate. It controls a number of weeds in lawns. The lawn may appear discoloured for a few days after treatment, but will recover quickly.

MCPA is available in mixtures with mecoprop or dichlorprop to control lawn weeds.

Mecoprop is used in mixtures with 2,4-D or MCPA to control White Clover, Black Medick and Lesser Trefoil in lawns. It is also available with ioxynil to control lawn weeds including Slender Speedwell.

Oxadiazon is not yet available to gardeners, but may become so. It is mentioned because it is effective against the Pink-flowered Oxalis, for which there is no other treatment. It is also effective against Field Bindweed.

Paraquat/diquat is a herbicide mixture for giving a quick kill of existing weeds. It does not persist and treated ground can be used quite quickly. It has no lasting effect on perennials as it does not kill their roots. This is a poisonous chemical, but is safe if used as directed. Use only the products designed for gardens.

Propachlor is used for killing annual weeds as they germinate among various flowers and certain vegetables. It must be applied to weed-free soil. It persists for up to about two months.

Simazine is a useful herbicide in gardens. For paths and drives, it is available alone or mixed with other herbicides such as aminotriazole, MCPA or paraquat/diquat. These mixtures should not be used elsewhere. The added chemicals kill existing weeds and the simazine prevents any more from becoming established. Simazine is also available alone or with 2,4-DES to give long-term control of germinating annual weeds in shrubberies, rose beds and under fruit bushes and trees. It persists throughout all the season.

Sodium chlorate is the traditional weedkiller. It is not always very suitable for the small garden, because it can move through the soil, especially down slopes, and so damage shrubs and trees at a distance. It is very effective in killing large and deep-rooted weeds. It persists in the soil for between three and twelve months depending on the amount applied. Clothing, sacking and plant remains if soaked in it are liable when dry to burst into flames spontaneously. Most products now contain a fire depressant to reduce this risk.

2,4,5-T is available either alone or with 2,4-D to control many large and deep-rooted weeds such as Brambles, Stinging Nettles, Thistles and woody plants, for which 2,4,5-T is very effective.

Tar oil is available for control of lawn weeds, particularly Slender Speedwell and mosses. It is, however, liable to scorch the grass. It can also be used on paths and drives. It is a poisonous chemical, but is safe if used as directed.

Further information on herbicides is given in Chapter 6 (page 81).

OTHER SOURCES OF INFORMATION

Reference can also be made to the *Directory of Garden Chemicals* prepared by the British Agrochemicals Association. This is a useful guide to the 200 or so garden chemicals at present available to gardeners. It includes all pesticides as well as herbicides (i.e. those for insects, fungi, slugs, etc.) and gives their trade names. Also listed are the Company Members of the Association who supply garden chemicals, their addresses and the names of key personnel within the companies and their positions. The *Directory* is available from The British Agrochemicals Association, Alembic House, 93 Albert Embankment, London SE1 7TU.

Further information can be obtained from the *Weed Control Handbook* issued by the Crop Protection Council, edited by J. D. Fryer and R. J. Makepeace and published by Blackwell Scientific Publications. This handbook is revised periodically. It contains a wealth of detailed technical information on weeds and weed control in all situations. The two volumes are obtainable through any bookseller.

There is also a small leaflet entitled *Chemical Weed Control in Your Garden* available for a small charge from the Librarian of the A.R.C. Weed Research Organisation, Begbroke Hill, Yarnton, Oxford, OX5 1PF.

GLOSSARY

Annual: a plant completing its life-cycle within twelve
months, although not necessarily in the same calendar
year.

Axil: the angle between a leaf and the stem on which
it is growing.

Biennial: a plant completing its life-cycle in two years,
but not flowering in the first.

Bract: a small leaf-life structure associated with flowers.

Bulbil: a small bulb-like structure.

Dormant: a live seed that fails to germinate under ideal
conditions is said to be dormant.

Floret: a small flower.

Habitat: the type of place in which a plant grows, e.g.
paths, lawns, etc.

Herbicide: a chemical capable of killing plants.

Node: a point on a stem where leaves occur, the leaves
may be reduced to scales on underground stems.

Perennial: a plant living for longer than two years.

Procumbent: wholly prostrate on the ground.

Stamen: the pollen-producing male part of a flower,
usually consisting of pollen sacs on a stalk.

Tap root: the main root of a plant, which is thickened
in some weeds.

Tuber: a swollen underground stem or root.

Viable: a seed when alive is said to be viable.

INDEX